keeping well at work

Other 'TUC Guides' available from Kogan Page:

Your Rights at Work

2nd Edition

'This book should have a readership of about 25 million. Just about every worker in this country has something to gain from dipping into this simply written guide to their employment rights.'
The Observer

This practical guide, written by employment experts at the TUC, has been fully updated and expanded to include information on gains in employment law, pensions and new issue such as e-mail privacy. Your Rights at Work offers comprehensive, jargon-free advice both for union and non-union members. It covers sensitive issues such as:

- parental leave and maternity rights;
- discrimination and bullying;
- workplace monitoring;
- dismissal and redundancy;
- pay and holiday rights;
- employment tribunals.

Planning Your Pension

Sue Ward

How can we ensure that the pension planning we put in place now will serve us well in retirement? What are our pension rights and what choices are on offer?

In this practical jargon-free guide for everyone – trade unionists and non-trade unionists alike – Sue Ward answers these questions and many more. She provides a much-needed layperson's guide to the complexities of pension provision and the various options available to us. Packed with sound impartial advice and lots of real-life examples, Planning Your Pension guides you through the pensions minefield to give you a clear insight into:

- the different types of pension scheme;
- contracting out of SERPs and the new S2Ps;
- how to change your scheme;
- how to negotiate your package;
- your employer's role;
- the legal rules;
- what to do if things go wrong.

Both books are available from all good bookshops. To obtain further information please contact the publisher at the address below:

Kogan Page Limited
120 Pentonville Road
London N1 9JN
Tel: 020 7278 0433
Fax: 020 7837 6348
www.kogan-page.co.uk

keeping well at work

2ND EDITION

a TUC guide

worksmart
www.worksmart.org.uk

philip pearson

KOGAN PAGE

Dedicated to the memory of Dr Ruth Elliott

First published in 2001
Second edition 2004

Kogan Page Ltd
120 Pentonville Road
London N1 9JN
www.kogan-page.co.uk

British Library Cataloguing in Publication Data

A CIP record for this book is available from the British Library

ISBN 0 7494 4152 6

Typeset by Saxon Graphics Ltd, Derby
Printed and bound in Great Britain by Cambrian Printers Ltd, Aberystwyth, Wales

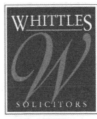

Contents

Foreword

Every year around 2 million people in Britain are injured or made ill by their work. Most make a full recovery, although every week 500 people have to give up paid work altogether because of a work-related illness or injury. And of course these figures only include those who report their injuries or illness. Many other people will suffer in silence, or experience discomfort that, although real enough, is not severe enough to make them seek medical assistance.

There was a time when talking of work-related health problems would conjure up images of heavy industry, dangerous chemicals or traditionally dangerous jobs such as construction, mining or deep-sea fishing. These problems have not gone away, and still require constant vigilance. But there are new workplace epidemics such as stress, and particular problems caused by jobs in newer parts of the economy such as call centres, offices and supermarkets. Issues such as bullying, harassment and violence that have probably always been with us are also now recognized as health problems. And some problems, like back pain, can affect every type of job.

The title of this book has been chosen very carefully. We want people to stay well, and avoid illness and injury. The way to do this is know what the risks and hazards are in your workplace and how to stop them affecting you and your workmates. Knowledge is power when it comes to keeping well at work.

Nor do we take the view that it is all up to your employer. Of course the employer should take every sensible precaution to make your workplace safe, and follow health and safety legal obligations and recognized good practice. Indeed we spell these out in this book. But we all have an obligation to work safely and to take responsibility for looking after ourselves as much as possible.

So while, as you would expect in a TUC guide, we carefully set out your legal rights, we put as much emphasis on prevention and how to recover from any problems and get back to work as soon as possible. There is advice on how to get the best from your GP and tips on other sources of help and assistance. In particular, since the first edition of this book the TUC has launched workSMART – our world of work Web site, which includes a substantial and constantly updated section on keeping well at work at www.worksmart.org.uk/health.

Of course people should get compensation for illness or injury when their employer (or someone else) is clearly to blame. Indeed Britain's unions regularly take cases and win more than £300 million a year for their members. But we wish we did not have to. Prevention is always better than compensation. Nor do we support the growth of the 'claims farm' industry – the heavily advertised opportunities to extract compensation seemingly irrespective of the strength of the claim. Our experience is that many have failed to obtain the riches promised, and the net effect has been to undermine this whole area of law and hinder those with serious cases, such as those brought by unions or expert personal injury lawyers.

This book is for everyone at work, not just trade union members. But inevitably there are frequent references to unions and to union health and safety representatives – the 200,000 workplace reps who make sure that unionized workplaces are safer and healthier than non-union workplaces. Too often they are unsung heroes, and do not get the credit they deserve. Yet the figures speak for themselves. Unionized workplaces have fewer than half the number of major injuries of non-unionized workplaces.

Unions have an important role in health and safety in the UK. This is because health and safety standards are set by the Health and Safety Commission (HSC), which brings together employers, unions and experts, and enforced by the Health and Safety Executive (HSE), which works to the HSC. Unions are always in the forefront of campaigns to improve standards and toughen up protection.

You can see that in our campaign for a new corporate killing law that would make directors responsible when company decisions are responsible for loss of life, as we have tragically seen in too many recent fatalities. The government has agreed in principle to this, and we have every expectation of an early change in the law.

This would be another union contribution to making Britain a safer place to work. Indeed Britain is already one of the least dangerous places to go to work, but every work-related illness or injury is unnecessary and we see this book as another union contribution to achieving the government's ambitious targets to reduce workplace illness and injury further.

But most of all this book has a simple purpose – to keep you healthy by setting out your rights; the steps you can take to make your workplace safer and a better place to work; and where you can get help. Use it!

Brendan Barber
General Secretary of the TUC

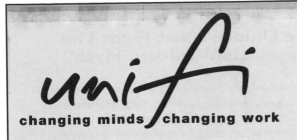

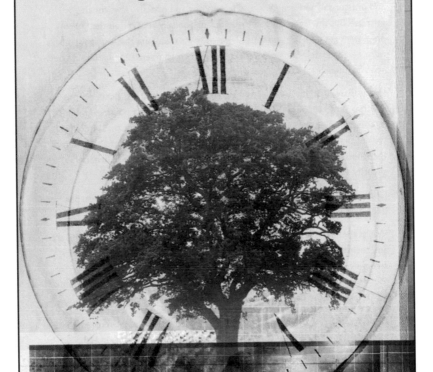

compensation as the reason. They have in effect blamed the victims when demanding that "reforms" to the industrial accident and disease compensation system.

What Trade Unionists must know is that this campaign is based on untruths. The amount of personal injury litigation in the UK is not high by international standards and the number of claims have in fact decreased.

The most recent statistics from the Compensation Recovery Unit of the Department of Work and Pensions, with whom insurance companies must register all personal injury compensation claims, show the degree to which the "compensation culture" is a myth.

Between 2000/2001 and 2002/2003 the number of disease claims fell by 26 percent, whilst employers liability claims went down by 16 percent.

Moreover, the UK comes at the bottom of the league table of compensation costs within industrialised nations. A mere 0.6 per cent of UK's GDP can be attributed to the costs of obtaining compensation for those who have suffered accidents or industrial disease. In Germany the figure is 1.3 percent; in Italy it is 1.7 percent. In the USA the level is 1.9 percent – more than three times that of the UK.

The truth is that insurance companies have not been overwhelmed with increased compensation payments to Trade Union Members. Their profits have been hit by the dramatic decline in the stock market over the last few years. They are using the myth of the 'compensation culture' to try to change the compensation framework to limit access to justice for accident and industrial disease victims and to reduce the levels of compensation they are required to pay. If they succeed workers will in effect be paying for the decreased value of the insurance companies' stock market investments.

What the insurance industry should be made to do, by regulation if necessary, is to reward companies with good health and safety records. This would help protect employees and in the long run reduce the number of claims to which they are exposed.

The Trade Union Movement must defend its members' right to seek compensation when they have been injured by negligent employers. The fact is that many are not recovering enough compensation for the injuries and diseases from which they suffer, with the level of damages awarded by UK Courts remaining low by international standards. Trade Union Members must know that the so called tidal wave of British compensation claims has not happened. At the moment a steady stream would be a better analogy.

Acknowledgements

For this second edition, I would like especially to thank Becky Allen, Peter Purton and Stuart Gillings for their advice in bringing *Keeping Well at Work* up to date.

I am very grateful to the many individual workers who gave up their time to be interviewed, sometimes more than once, commented on drafts and offered their advice. Their experiences are reported here as accurately as I was able, as separate chapters or to illustrate a particular point. My thanks also to their union reps and full-time officers, who provided further material and checked drafts. To protect their identity, they appear under pseudonyms. Certain case details have been altered, but not, I hope, the essence of their stories. I trust that the narratives do justice to your experience: Gita, Eddie, Louise, Toni, Teresa, Mark, Dougie, Dave, Cath, John, Gill, Mary, Kevin, George, Mike, Janet, Christine and Ann.

For specialist advice and information on major and minor issues, comments on drafts, suggestions on sources, I greatly appreciate the assistance I received from the following individuals and organizations: Diana Lamplugh, OBE and The Suzy Lamplugh Trust; the Telephone Helplines Association; the RNIB; Stuart Gillings, partner at Bolt Burdon, solicitors, for his advice on personal injury claims; the Labour Research Department; the National Hazards Campaign; Julie McLean, speech and language therapist, and member of the Voice Care Network; Dr Claire Highton; Christine Baker and Chris Ryan of the National Association for Premenstrual Syndrome; Noreen Tehrani; Dr David Holman; Marian Bell; Elaine Hislop, Islington Law Centre; Alan Denbigh; James Royston; Neil Budworth, and Dr Linda Grant.

I have relied a great deal on the knowledge and experience of workplace health and safety from trade unionists who are experts in their fields. Union initiatives are helping to accelerate the pace of change in the way Britain understands and tackles occupational health. I make extensive use of the wide range of published guidance produced by their organizations. In particular, I would like to thank Owen Tudor, Nigel Stanley, Hugh Robertson, Hope Daley, Cath Noonan, Sara Marsden, Dave Turnbull, Peter Purton, Doug Russell, Alan Scott and Kim Sunley. For insights into the work of community-based occupational health and hazards projects, and the innovative work they are undertaking to help empower individuals faced with work-related ill health and injury, my thanks to Mick Holder, Simon Pickvance, Sharon Cave, Mick Williams, John Bamford, Jim McCourt, Jill Barlow, Stefan Harper and many others in the Hazards network. I am also grateful for the support and advice of fellow workers and colleagues at Industrial Relations Services, particularly Howard Fidderman, Adam Geldman, Dave Martin, David Fox, Dr John Ballard, Becky Allen and Peter McGeer. And finally, for their support and encouragement, my thanks to Nony, Anya and Aidan, and to Andy, Linda, Francesca, Mick, Alison, John, Neil and Jill.

Introduction

This handbook, *Keeping Well at Work,* aims to make you better informed about the most common 'modern' health risks at work, and how you can tackle them. From asthma to zero tolerance, it offers positive suggestions about preventing ill health and improving your well-being at work.

Only connect! Work and ill health

All too often, people don't make the connection between their job and their health. We have interviewed workers in apparently 'safe' jobs – a checkout operator, a hotel receptionist – suffering from severe work-related ill health and injury. But they became ill without relating their health to their work.

Perhaps there are three reasons for this. First, a kind of 'denial' can set in, as you subconsciously adapt to the stresses you are working under. Second, when you visit your GP, you assume he or she knows all there is to know about work-related ill health and injury. Unfortunately, this may not be the case. Although seeing your GP is often essential, a hard-pressed GP, perhaps untrained in 'occupational health', may miss some vital signs. And third, while union-elected safety reps advise and support many thousands of workers each year, more than half of Britain's private sector workplaces aren't unionized, and do not have workplace safety reps.

How to use this handbook

We approach each of the work-related health issues in this handbook in the following way:

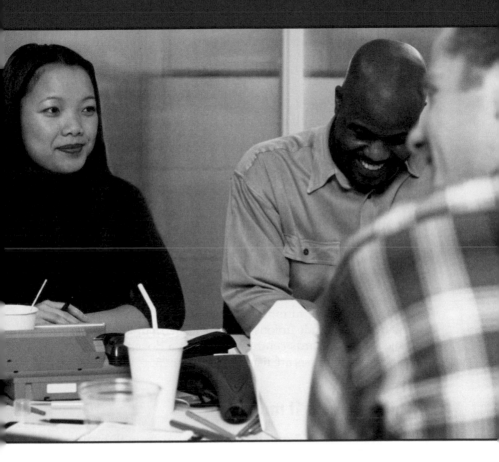

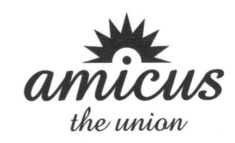

- Through **real-life case studies**, based on interviews and other documentation, featured anonymously in most instances.
- With **information on the scale of the problem**, the key symptoms and causes of each health issue.
- With **information on employer's legal duties** to make work safe.
- With **good practice advice** on how unions, employers and voluntary organizations tackle health and safety issues.
- By offering **self-help ideas**: what you can do for yourself and where to seek help.
- By providing **'hard' legal advice**, especially if you are faced with a serious health problem.
- With **further information sources**: at the end of each chapter we refer to advice handbooks, guides and leaflets, published by unions, voluntary organizations and employers. Many of these are also available on Web sites. If you don't have access to a computer at home or work, it may well be worth spending an hour at an Internet café, downloading the information available. Good examples of Internet resources are the excellent information on disability at work, available at www.disability.gov.uk, and workSMART, the TUC's health at work Web site: www.worksmart.org.uk.
- With the **helpline numbers** provided for national voluntary organizations, trade unions, and the local occupational health projects and hazards centres in around a dozen cities in the UK. Union helplines are listed in Chapter 25.

Two million, or is it more?

Every year, more than 2 million people suffer from ill health or injury caused, or made worse, by the job they do. Official figures from the Government's Health and Safety Executive (HSE) show that the most common forms of work-related ill health are back problems and other disorders affecting the neck, shoulders, arms, hands, wrists and fingers. The soft tissue damage involved, to muscles, tendons and nerve supply, lend to these injuries the name: *musculo-skeletal disorders*. They affect over 1 million people a year.

These injuries can happen suddenly – a tripping accident at work, a pulled muscle due to poor lifting technique. Or they creep into your system as 'slow accidents', through the daily 'wear and tear' of repeating the same actions without adequate work variety or rest breaks, and often under pressure.

Yet psychological stress is now seen as *the* major workplace health issue. TUC polls of union health and safety representatives reveal that stress, anxiety and depression, linked to excessive workloads, have become their number one concern. The main causes cited are heavy workloads, staff cuts and the 'long hours culture'. And bullying is now recognized as a major 'trigger' of stress-related ill health.

A truer picture

Latest official figures show that 2.3 million people believed they were suffering from an illness caused, or made worse, by their current or previous work (see Table 0.1). Yet it is hard to escape the conclusion that official figures grossly underestimate the extent of work-related ill health:

- An estimated 5 million workers suffer from high levels of stress at work, according to an HSE-funded study of 8,000 people.
- One in 10 people were bullied at work over the past six months, or 2 million of the UK's 24 million workers.
- Over 1.3 million violent incidents were committed by members of the public against workers in 1999, according to Home Office figures.
- About 800,000 women experience 'severe' discomfort from premenstrual syndrome, but their condition is largely unrecognized at work.

A key reason for the 'reality gap' between the special studies and the general official figures has been hinted at already: people don't necessarily see the link between their ill health and their job, until someone asks them.

Table 0.1 Work-related ill health in Britain, 2001 / 02

Types of Work-Related Illness	**Number of Workers Affected**
Musculo-skeletal disorders: strains and pains in the back, upper and lower limbs, hands and fingers	1,126,000
Stress, depression or anxiety	563,000
Breathing and lung problems, including asthma	168,000
Hearing problems, including deafness and tinnitus	87,000
Heart disease, heart attack or other circulatory system problem	80,000
Headache and / or eyestrain	54,000
Skin problems	39,000
Infectious diseases	33,000
Other complaints	171,000
Total	**2,321,000**

(Source: Health and Safety Executive, *Self-Reported Work-Related Illness in 2001/02*)

Keeping Well At Work offers positive, practical advice about preventing ill health and injury, and improving well-being at work.

The author and the TUC have done their best to make sure this book is accurate at the time of writing, but inevitably have had to simplify issues. You should not rely on the general advice provided by this book as a substitute for detailed legal or medical advice on your own position. You should always take expert advice if necessary.

Part 1

Health and safety law

7 Chemical cocktail

Eddie's story

Eddie's symptoms felt like the flu, except that, unusually, he ached only from the waist upwards, not all over. A keen cyclist, who regularly worked out in a local gym, he considers himself to have been healthy and active, until a sudden and inexplicable deterioration in his health.

Eddie was working as a receptionist in a central London hotel. He visited his GP, who referred him to a chest consultant. He underwent a series of breathing and lung function tests, which were inconclusive, and was prescribed medication as a precaution against pulmonary embolism (blood clotting).

Coincidentally, Eddie was transferred to back-office work for some months and, working well away from the reception area, he noticed that his health began to recover. At the next appointment with the consultant, he described the changes to his condition. The specialist asked whether he had been in contact with any chemical or other substances that might have contributed to the illness.

Shortly afterwards, Eddie was transferred back to reception, where the symptoms began to recur. He started to keep a diary of his symptoms: a very sore throat, occasionally accompanied by white blisters inside the throat; a burning sensation of the throat; and sore eyes that would crust up overnight.

Chemical exposure?

He felt a tingling sensation in the tongue, a warning signal, on one occasion when cleaners were mopping the floor and using a pine-scented aerosol to freshen the air. It then occurred to Eddie that these may be linked to his ill health. So he met the hotel's health and safety officer, who unfortunately expressed little interest, and appeared to be more concerned about the cost of changing the stock of cleaning materials.

Eddie now became very anxious about how to deal with his condition, and decided, after advice from a colleague, to fill in his first accident report. As a result, the health and safety manager banned the use of the air freshener and the particular brand of floor cleaner.

However, the manager's instructions for controlling the use of the chemical cleaners and spray were not taken seriously. The materials were not disposed of. The contract cleaning staff continued to use them quite frequently, perhaps because of a high staff turnover, or through negligence on the part of their supervisor. Over the next few months, Eddie's symptoms of a sore throat, tongue and mouth recurred on several occasions. On one occasion, he discovered that the same aerosol was being used, and on another, the same cleaning fluid. The supervisor denied responsibility. A further bout of ill health appeared to be linked to the use of a new floor cleaning fluid on trial in the hotel, which no one thought to tell him about. Each time, he recorded the illnesses in the accident book.

For the rest of the year, Eddie was temporarily redeployed to another office where, although still not returning to anything like his former good health, he suffered no further health setbacks. Some months later, he returned to the reception area, where he suffered the most serious attack to date. Later, it emerged that, before he had arrived at work, the night cleaners had sprayed the whole reception and front office with the banned pine-scented air freshener. The symptoms returned with a vengeance soon after he arrived at work, including a bright red rash all over the exposed areas of his neck, arms and throat. The inside of his throat became extremely sore.

This time round, Eddie became seriously unwell, and also began to suffer from depression. He felt that no one, apart from a work colleague, was listening. This time, he was off work sick for a week and, as a result, because his absence lasted more than three days, the hotel was legally required to submit an incident report to the council's environmental health department (see RIDDOR Regulations, page 38).

Until that occasion, not one of the periods of absence had been reported to the local authority, because, in-line with RIDDOR rules, they lasted less than three days. A safety inspector visited the hotel.

Now suffering from depression as well as the persistent effects of the most recent chemical exposure, Eddie was unable to work for more than three months. His GP prescribed antidepressants, and he began to see a counsellor. When he returned to work, his employer transferred him to an alternative job, away from reception, but on a lower salary.

The final exposure had even wider consequences. He became highly sensitized, so that he developed unexpected allergic reactions to a wide range of domestic chemicals, sprays, cleaning materials, detergents, certain perfumes, air fresheners and deodorants. Some washing-up liquids made his mouth swell up inside.

He has since learnt that another office worker has described exactly the same kinds of symptoms. Management installed an extractor fan in her office, although it does not appear to be making any real difference. The management had not appreciated that because only one or two workers were affected this did not mean that their condition was not due to a hazardous exposure at work.

Despite extensive tests, including 'patch' skin tests at another hospital, Eddie is no closer to a full diagnosis of his condition.

A 'negligence' claim

Eddie is considering a personal injury negligence claim against the hotel. This would take into account:

- A possible breach of the employer's duty of care.
- Possible multiple breaches of employer's duties under the Control of Substances Hazardous to Health (COSHH) Regulations 2002. The hotel's health and safety manager failed to take adequate measures to identify and control substances suspected to be harmful. His instructions to the cleaning supervisor, to substitute safer cleaning materials, were not followed. The high turnover of contract cleaners

made this situation worse: they were unaware of the manager's instructions. By failing to control the risks, the hotel had not provided a safe place of work and a safe system of work.

- Loss of earnings: because he was transferred to a lower-status job.
- Pain and suffering: repeated episodes of physical ill health and psychological stress.

Explanations

There are three possible explanations for Eddie's condition. None are well understood. The first two are types of 'dermatitis', or 'inflammation of the skin'.

- **Allergic contact dermatitis** is caused by certain chemicals, called 'sensitizers'. The body's immune system reacts to one of these substances. In many cases, people can work with, or be exposed to, a substance for years without suffering an adverse reaction. They then suddenly develop a contact reaction that develops into dermatitis. Once a person becomes sensitized by exposure to one substance, even minor exposures may cause a severe reaction, and the person may also become cross-sensitized to a range of substances. Rubber latex, found in gloves and face masks, is a common cause of skin sensitization.
- **Irritant contact dermatitis**, which arises from working with substances (eg cleaning fluids) that physically damage the skin when they come into contact with it. Examples include acids and alkalis, such as caustic soda or cement, or organic solvents, such as white spirit. Occupational dermatitis is commonplace among cleaners, catering staff (especially chefs), hospital workers, especially nurses and laboratory staff, and some craft occupations.
- **Multiple chemical sensitivity**, stemming from exposure to a 'cocktail' of harmful substances. Symptoms of multiple chemical sensitivity vary widely, often involving the upper

respiratory tract (blocked nose, burning sensation), muscle or joint pains, headaches / tiredness, and central nervous system complaints (eg memory problems, insomnia). Occupational health specialist Becky Allen suggests that 'the initial cause will usually have been exposure to one of a small number of *initiators*', such as pesticides, solvents or formaldehyde. Once sensitized, an individual's symptoms may appear after exposure to a much wider range of chemical 'triggers', ranging from air fresheners and alcohol to cosmetics, newsprint and tobacco smoke.

The causes of chemical sensitivity are not well understood, but are perhaps best described as a disproportionate response of the immune system in some individuals. It is important that sufferers are aware of the limitations of scientific understanding when they are seeking help and advice from health specialists.

Hazardous substances: the law

As many as 25,000 workers each year fall ill as a result of exposure to substances that are hazardous to health. Yet health and safety law requires employers to control exposure to hazardous substances to prevent ill health occurring. They also have a further duty to report to health and safety inspectors certain types of ill health and injury arising in connection with hazardous substances. Employers that fail to comply with these laws are liable to prosecution by a health inspector. Meanwhile, injured employees can also make a civil claim for damages against their employer.

In response to public anxiety, the Health and Safety Executive (HSE) strengthened its powers over **hazardous substances** at work with the Control of Substances Hazardous to Health (COSHH) Regulations 2002. The Regulations also bring new controls over **occupational asthma** and **asbestos**. Exactly how dangerous a chemical is depends on several things:

- What the chemical is and whether it is solid, liquid or gas. Some chemicals damage your skin. Some chemicals, as fumes and vapour, can be breathed in and damage your lungs and other organs. And some chemicals are absorbed into your body through your skin, or when you swallow, and damage your organs, such as your brain, heart, liver and kidneys. Once in your body, some chemicals can also affect your fertility and your unborn children. High levels of exposure to some chemicals and toxins can cause psychiatric ill health, such as depression, irritability, personality changes, memory loss, sleep loss and sexual problems.

- The length of time you are exposed to the chemical.
- The amount of the chemical you are exposed to.
- Some chemicals are also dangerous because they are flammable or explosive.

Read the label

You can tell if chemicals you work with are dangerous just by reading the label. Under the Chemicals (Hazard Information, Packaging and Supply) Regulations (CHIP), chemicals should be delivered and stored in properly labelled containers. The label gives you important information about the hazards of the substance, for example if it is toxic or likely to cause skin burns or allergic reactions. The Health and Safety Executive's leaflet *Read the Label: How to find out if chemicals are dangerous* explains what the symbols on chemical labels mean.

It is vital that you *read the safety data sheet* that should accompany the product. A safety data sheet gives you more detail about a substance's hazards and the precautions you need to take. Your employer should keep copies of safety data sheets for the products you use, and should tell you how to use the products safely.

Who is at risk?

Occupations most at risk from exposure to hazardous substances are hairdressing, repetitive assembly, nursing, construction and farming. But a wide range of other occupations may expose you to harmful chemicals:

- building work, heating engineering and plumbing, because of the asbestos remaining in many buildings;
- vehicle maintenance, because of asbestos dust in brake linings;
- catering, because of contact with soaps and detergents, and regular handling of foods such as flour, sugar, meat, fish and fruits;
- cleaning (both domestic and industrial), because of the use of cleaning products and sprays;
- painting and decorating, because paints and varnishes may contain toxic solvents;
- gardening, because it may involve the use of chemical pesticides;
- printing, because some inks, developers, oils and solvents are harmful to your skin;
- hairdressing, as hair dyes and perming solutions contain ammonia and other harmful chemicals;
- dry cleaning, as this involves the use of toxic solvents;
- nursing and healthcare, because of the widespread exposure to biological agents, such as pathogens (bacteria or viruses).

Risk assessment – the Regulations

The COSHH Regulations require employers to carry out a risk assessment and decide what precautions to take, either to prevent or adequately to control exposure. The Regulations require employers to use a 'hierarchy of controls'. This means that your employer should:

- change work procedures to eliminate the chemical completely;
- substitute the chemical with a safer one;
- if using a hazardous chemical cannot be avoided, protect workers from exposure to it by enclosing the process;
- provide adequate ventilation.

Only when these steps have been taken should your employer rely on personal protective equipment, such as gloves and masks, to prevent your exposure.

The Regulations also set upper limits on the amount workers should be exposed to certain chemicals. Details of these 'occupational exposure limits' are published on the HSE's Web site at: www.hse.gov.uk.

Employers also have special legal duties to protect new or expectant mothers from exposure to biological hazards, or to certain chemicals, such as lead and mercury, and from handling drugs or pesticides – see the HSE's Web site and *A Guide for New and Expectant Mothers who Work*, available at: www.hse.gov.uk/mothers. If in doubt, contact your GP (see page 230).

Avoiding asthma

Each year, around 3,000 workers develop asthma as a result of the job they do. Asthma is a condition that affects the airways of the lungs. When these airways become inflamed, they are narrower and carry less air to the lungs. Wheezing, coughing, shortness of breath and a tight chest are common symptoms.

A wide range of substances can cause occupational asthma, including dust from flour, grain and wood; isocyanates (chemicals found in two-pack spray paints); glutaraldehyde (a disinfectant mainly used in the healthcare sector); and natural rubber latex. Severe skin irritation and respiratory problems can result from wearing latex-powdered gloves, still widely used in healthcare. It is essential to discuss your concerns with your GP – see our case study on page 241.

In response to growing concern about this disease, including worries frequently raised by unions, the HSE included an Approved Code of Practice (ACoP) on Asthma in its COSHH Regulations 2002. Employers must apply the same risk assessment approach as set out in the COSHH Regulations. As a general rule, exposure to substances with the potential to cause asthma should be prevented. If that is not possible, employers must do all they reasonably can to control exposure. The ACoP also says that employers must:

- ensure their risk assessments take into account how seriously ill a worker could become if they failed to control exposure;
- arrange 'health surveillance' (regular health checks) by an occupational health professional for workers who may be exposed to risk;
- report new cases of occupational asthma to a health and safety inspector.

Killer asbestos

Asbestos dust is highly dangerous. Official figures show that asbestos-related diseases (mesothelioma, asbestosis and lung cancer) kill at least 3,000 people a year, more people than any other single work-related cause. The government banned its use in new buildings and products from 1999 after a long campaign by trade unions, the TUC, local safety campaigns and *Hazards* magazine.

But there is still a great deal of asbestos in older buildings. Stripping it out is high risk. Breathing air containing asbestos dust can lead to potentially fatal asbestos-related diseases.

Many of today's asbestos victims worked in building trades, as carpenters, joiners, shopfitters, plumbers, electricians and demolition workers. They were exposed to asbestos dust in their day-to-day work with asbestos materials, or because work with asbestos was carried out near them. Many have worked for more than one employer where asbestos is present on site.

If you carry out any repair or maintenance work in buildings with asbestos materials, you could be exposed to asbestos dust and breathe it in without realizing it. Although you may be exposed only to small quantities, if this is repeated often it can build up in your lungs. You could develop an asbestos-related disease in later years.

Who is responsible for asbestos?

Anyone with responsibility for a building containing asbestos has a 'duty to manage' the asbestos, under the Control of Asbestos at Work (CAW) Regulations 2002 (CAWR). Regulation 4 of the CAW

Regulations takes effect from May 2004. By then, any duty holder is required under the regulations to find asbestos-containing materials in the premises, check its condition and tell anyone likely to come into contact with it during building works.

Employers' responsibilities are spelt out in an Approved Code of Practice (ACoP), *The Management of Asbestos in Non-Domestic Premises*. The ACoP is published with the main Regulations, and requires employers to carry out an asbestos risk assessment that must:

- identify the type of asbestos;
- identify the nature and level of your likely exposure;
- aim to eliminate or reduce any exposure to risk;
- include arrangements to monitor exposure.

Do not start work until this risk assessment has been carried out. Stop any job immediately if you suspect asbestos is present in any material. You must be given a copy of the work plan your employer is required to produce. You should not work with any asbestos product unless your employer is a licensed asbestos remover. You should not work with any asbestos product unless you have received adequate information, training and protection from your employer.

Safer work with asbestos

Never work with asbestos without a full risk assessment and safe plan of work. The following are examples of safer working practices:

- Keep asbestos material damp.
- Don't use power tools, as they create dust.
- Use personal protective equipment given to you. This may include a suitable mask and disposable overalls.
- Make sure you are properly trained to use the mask.
- Clear up waste, and put it in suitable sealed containers.
- Don't use brooms or brushes; use a suitable vacuum cleaner.

- Wash your hands and face before eating, drinking and smoking.
- Don't take overalls home. Your boss should wash them for you.

Advice and support

Asbestos is a major issue for trade unions and community-based safety campaigns. If you are in any way unhappy about working with asbestos, stop the job to avoid further exposure. Speak to your site safety rep, shop steward or full-time officer. Contact a local asbestos campaign or support group – many local projects have taken up the asbestos issue in their community. Details of a local asbestos campaign near you are available on the *Hazards* magazine Web site: www.hazards.org/organisations.htm.

Claiming compensation?

If you are suffering from ill health as a result of exposure to asbestos dust, even if your exposure occurred while working for more than one employer, you may be able to make a personal injury claim – see Chapter 24. But you will need specialist legal advice, whether or not it is settled out of court. You must bring your claim by starting legal proceedings within three years of the date of your illness, so do not delay any further if you believe you may have a claim and have not made one so far.

Further information

Asthma at Work: Are you eligible for compensation?, available from the National Asthma Campaign (tel: 020 7226 2260; Web site: www.asthma.org.uk).

Caring for Cleaning Staff, UNISON health and safety unit (tel: 020 7551 1446).

Chemical Hazards at Work, Allen, Becky, available from London Hazards Centre (tel: 020 7794 5999).

Control of Substances Hazardous to Health, USDAW (tel: 0161 224 2804; Web site: www.usdaw.org.uk).

Hazardous Substances at Work: A safety reps guide includes practical advice on the new COSHH Regulations, asthma and asbestos hazards. Available from Labour Research Department (tel: 020 7928 3649; Web site: www.lrd.org.uk), price £4.95.

Hazards magazine publishes *Asbestos Campaign Guide; Asbestos Fact Sheet; Construction Hazards*; and contact details of local asbestos campaigns (tel: 0114 267 8936; Web site: www.hazards.org.uk).

Health and Safety Essentials, Amicus.

HSE publications include: *Asbestos Dust: The hidden killer, are you at risk? A short guide to managing asbestos in premises; COSHH: A brief guide to the Regulations; Read the Label: How to find out if chemicals are dangerous; Preventing Dermatitis at Work: Advice for employers and employees; Occupational Dermatitis in the Catering and Food Industries*. Available from HSE Books (tel: 01787 881165; Web site: www.hse.gov.uk).

The Health and Safety Executive (HSE) asbestos Web page includes *Asbestos Campaign News* and other information: www.hse.gov.uk/campaigns/asbestos.

White Asbestos: It's still a killer, available from GMB general union (tel: 020 8947 3131; Web site: www.gmb.org.uk).

2 The right to keep well at work

You have the right to work in a safe and healthy environment. Employers have statutory duties to protect the 'health, safety and welfare' of their employees. Here, we set out these duties, and how you can enforce your health and safety rights, through consultation at work or through legal action.

Your employer's duties

The statutory protection of the Health and Safety at Work Act 1974 (HASAW Act), reinforced by Regulations and codes of practice, require your employer to provide a healthy and safe workplace. This means:

- a safe system of work;
- effective supervision of safety at work by a 'competent person'; and
- safe and competent people working alongside you – employers are liable for the conduct of their staff and managers.

Employers owe their workers a 'common law' duty of care. This means taking reasonable care for their safety, and not exposing them to unnecessary risks.

But things go wrong. An injury may happen suddenly and unexpectedly. Slips, trips and falls, or assaults and abuse from customers or clients, are among the main causes of serious injuries at work. But you may also suffer from one of the 'slow accidents' that take their time: repetitive work at a checkout, losing your voice in a classroom, or stress brought on by prolonged bullying.

Safety studies show that employees are twice as likely to be injured in a non-union workplace. In unionized workplaces, employers are required to work with union-appointed safety reps and set up safety committees (see box, page 38).

If you are faced with a serious health and safety issue, there are *three* main types of legal action open to you:

- **A personal injury claim**. If you suffer a work-related injury or ill health – either a *physical injury* or *psychologically*, as a result of stress, bullying or intimidation.
- **Quitting work because of intolerable conditions**. If you quit work to escape intolerable working conditions, you can claim unfair 'constructive dismissal' at an Employment Tribunal.
- **Reporting hazards to a safety inspector**. A safety inspector can take action against an employer for breaching health and safety Regulations (see page 25).

Yet most injuries and ill health at work are preventable, if employers carry out a proper risk assessment and take steps to deal with the risks they discover.

Here, in this chapter, we cover:

- health and safety law at work;
- health and safety Regulations;
- the difference between 'hazards' and 'risks';
- what you can do for yourself;
- taking legal action;
- protection from victimization;
- the 'six-pack' of Regulations;
- rights of union safety reps; and
- claiming industrial injury benefit.

Health and safety law at work

Your legal rights to health and safety at work are based on four main areas of law:

- **Health and safety legislation**. The Health and Safety at Work Act 1974 (HASAW Act) provides wide-ranging duties on employers to protect the 'health, safety and welfare' at work of all their employees. However, these duties are qualified with the words 'so far as is reasonably practicable'. This means that employers can argue that the costs of a particular safety measure are not justified by the reduction in risk that the measure would produce. Employers' safety obligations extend to other workers on site, eg contractors and members of the public.
- **Employment protection legislation**
  - *Unfair dismissal*. You are protected against unfair dismissal by the Employment Relations Act 1996, once you have completed one year's continuous employment.
  - *Claims for unfair 'constructive dismissal'* include situations where you can no longer tolerate your working conditions, for whatever reason.
  - *Protection against 'health and safety' dismissals* includes when you have left work believing you were in imminent danger (see page 33). There is no length-of-service requirement.
  - *Sex and race harassment*. Legal protection may be available in, for example, bullying cases.
  - *Whistle-blowing*. You are protected if, as a 'whistle-blower', you made what is called a 'protected disclosure' of health and safety information.
  - *Bullying*. You can take action against bullying (see Chapter 13), even though there is still no separate UK law on bullying itself.
  - *Disability*. You are protected if you have a disability covered by the Disability Discrimination Act (see Chapter 20).
- **'Common law' duties**. For employers, the main common law duties are:
  - *A duty of care*. Employers must take reasonable care for the safety of their workers, avoid exposing them to any unnecessary risks and ensure a safe system of working.
  - *Mutual trust and confidence*. This means employers must not, without reasonable and proper cause, conduct

themselves in a manner calculated or likely to destroy or seriously damage the relationship of trust and confidence between themselves and their employees. This duty obliges employers to ensure employees are treated with dignity at work, and to deal with any complaints fairly and seriously.

- **Your contract of employment and staff handbook**, including your employer's safety policy and specific safety procedures. These range from the use of display screen equipment and routines for working away from the office to anti-harassment procedures.

The Health and Safety at Work Act

The Health and Safety at Work Act (HASAW Act), which came into force in 1974 following sustained pressure from the TUC, was designed to overcome weaknesses of earlier health and safety law. It brought together under a single piece of legislation a broad framework of protection for all workers in all occupations.

The HASAW Act:

- placed new general duties on employers, ranging from providing a safe place of work to consulting with employees;
- created the Health and Safety Commission, and an inspectorate called the Health and Safety Executive;
- introduced new powers and penalties for enforcement; and
- placed *occupational safety* at the heart of policy and future Regulations.

Employers' duties

The HASAW Act defines a broad range of employers' duties, including:

- to provide a safe and healthy workplace, safe systems of work and safe plant and machinery;
- to carry out risk assessments, and take steps to eliminate or control risks;

- to inform employees fully about all potential hazards associated with any work process, chemical substance or activity, including providing instruction, training and supervision;
- to appoint a 'competent person' responsible for health and safety;
- to consult with workplace safety representatives and, if the union is recognized, to set up and attend a workplace safety committee; and
- to provide adequate facilities for employees' welfare at work.

The Health and Safety Commission

The Health and Safety Commission (HSC) is an independent body responsible for developing health and safety policy and good practice. The chairperson, currently Bill Callaghan, and its nine commissioners are Government appointees representing trade unions, employers and local authorities.

The HSC has overall responsibility for the control and development of occupational health and safety in Britain. The Health and Safety Executive (HSE), the Commission's 'enforcement arm', promotes safer working practices. Inspectors from the HSE and local councils together share responsibility for inspecting workplaces and enforcing health and safety laws.

Enforcement

The HASAW Act provides three main systems of enforcement:

- Improvement notices. These require an employer to take action to put things right within a specified time.
- Prohibition notices. These stop a hazardous operation if there is an immediate risk of serious personal injury to workers or to the general public.
- Criminal offences. These include an employer's failure to comply with an enforcement notice. Penalties include a fine or, in some cases, imprisonment. The HSE now 'names and shames' employers guilty of criminal offences on its Web site: www.hse-databases.co.uk/prosecutions/.

An inspector calls

Enforcing health and safety Regulations is split between local councils and the HSE. They enforce the same law, and their inspectors have the same powers. The only difference is that they inspect different premises.

Local council environmental health departments mainly inspect service sector premises:

- shops and stores;
- banks and building societies;
- most offices;
- hotels, restaurants and catering generally;
- sports and leisure;
- residential accommodation.

For the council phone number, consult your local telephone directory and ask for the environmental health service or environmental enforcement officer.

Health and Safety Executive inspectors cover most public sector workplaces:

- transport (rail, air, bus);
- health service, and other government offices;
- manufacturing industry;
- construction sites.

There is an HSE office in most major cities, and other local offices. For contact details, see Further information, page 40.

When visiting workplaces with recognized safety representatives, inspectors should keep them informed of issues under consideration. Yet local councils have cut the number of their enforcement officers in recent years. A recent report in *Hazards* magazine (**72**, December 2000) shows that each inspector has to look after an average of 1,000 premises.

Regulating health and safety

Since 1974, the HASAW Act has been reinforced by over 400 Regulations, over 50 Approved Codes of Practice (ACoPs) and a wide range of Guidance for employers:

- **Regulations** are legally binding, and usually follow proposals from the Health and Safety Commission, and approved by Parliament. They cover the *general management* of health and safety at work, *work processes* (eg manual handling, use of display screen equipment) and *specific standards* (eg exposure to chemical hazards).
- **Guidance** gives employers practical advice on complying with the law.
- **Approved Codes of Practice (ACoPs)** are guidance with specific legal standing. These deal with hazardous materials and safe working practices. One of the most recent ACoPs clarifies employers' duties on preventing work-related asthma. Employers who are prosecuted for a breach of health and safety law, who have not followed a code of practice, are likely to be found at fault by the courts.

The 'six-pack'...

The 'six-pack' is the shorthand title of the half-dozen most widely quoted health and safety Regulations (see box, page 35). Chief of these are the Management of Health and Safety at Work Regulations 1999, also known as the 'Management regs'. They came into force in 1993, and were updated in 1999. Regulation 3 of the Management regs places a legal duty on employers to carry out a risk assessment as a first step in ensuring a safe workplace. Risk assessment lies at the very heart of a modern approach to health and safety at work.

Other 'six-pack' Regulations cover specific issues: heating, lighting and ventilation at work; the safe use of computer screens and keyboards; handling heavy or awkward loads; rest breaks; and personal protective equipment.

… plus two

In addition to the 'six-pack', two further sets of Regulations provide essential protection at work: the Safety Representatives and Safety Committees Regulations 1977; and the Working Time Regulations 1998.

The Safety Representatives and Safety Committees Regulations 1977 set out the rights of workplace safety reps to represent their members, take part in joint safety consultations with their employer and 'investigate potential hazards and dangerous occurrences in the workplace' (see box, page 38).

The Working Time Regulations 1998 implement two European Community Directives on the organization of working time and the employment of young workers (under 18 years of age). Protections include:

- a contractual obligation on employers not to require a worker to work more than an average 48-hour week;
- four weeks' paid holiday;
- minimum daily rest periods of 11 hours, unless shiftworking is involved;
- 20-minute daily rest breaks after six hours' work, with young workers entitled to 45 minutes if more than four and a half hours are worked; and
- a weekly rest period of 24 hours every seven days.

These basic limits on the working week make a vital contribution to health and safety at work. They are subject to detailed interpretation. Employers have the right to ask their staff to enter into a written agreement to opt out of the 48-hour limit, for a specific period or indefinitely. However, a worker is entitled to bring the agreement to an end without the employer's consent.

Other Regulations include specific **individual rights** at work, for example:

- the right to an eye test, if you are a regular user of display screen equipment;
- training in safe lifting and carrying;

- the right to receive full information on any risks you may face at work.

'Hazards' and 'risks'

The distinction between 'hazards' and 'risks' is crucial in the context of health and safety at work. A **hazard** is something that can cause harm, eg electricity, chemicals, working up a ladder, noise, a keyboard, a bully at work, stress. A **risk** is the chance, high or low, that any hazard will actually cause somebody harm.

Working alone away from your office can be a hazard, and the risk of personal danger may be high. Electric cabling is a hazard; exposed wiring places it in a 'high-risk' category.

This is a common classification of hazards:

- **Physical**: lifting, awkward postures, slips and trips, noise, dust, machinery, computer equipment.
- **Mental**: excess workload, long hours, working with high-need clients, bullying. These are also called 'psychosocial' hazards, affecting mental health *and* occuring within working relationships.
- **Chemical**: asbestos, cleaning fluids, aerosols.
- **Biological**: including tuberculosis, hepatitis and other infectious diseases faced by healthcare workers, home care staff and other healthcare professionals.

Vulnerable workers

In December 1999, the Management regs (The Management of Health and Safety at Work Regulations) were updated to include new protection for **new and expectant mothers** and **young workers**.

Where there are women of childbearing age in the workforce, any risk assessment must take account of how hazards might affect their health and safety (Regulation 16). As the Labour Research Department's guide points out, this includes exposure to any chemical agent, work process or working situation that

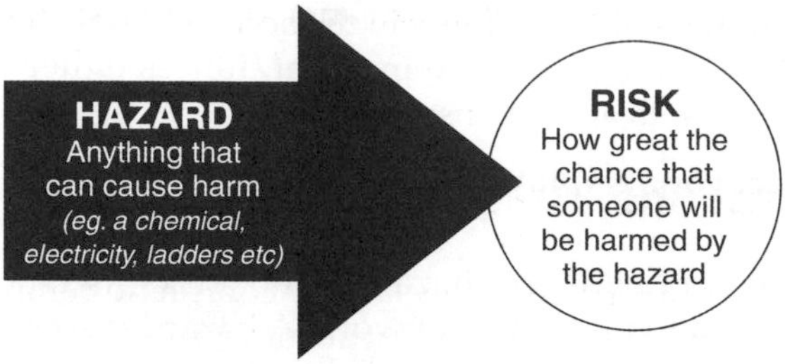

Figure 2.1 Hazard and risk
(Source: *GMB Guide to Risk Assessment*)

might damage the woman or her baby. Employers must take action to control or remove the hazard, or (for a pregnant employee) 'suspend the employee on full pay until the beginning of maternity leave'.

Employers must protect young workers, aged between 16 and 18 years, from risks to their health and safety arising from their lack of experience or maturity (Regulation 19). This debars employers from taking on young people for work that:

- is beyond their 'physical or psychological capacity';
- involves harmful exposure to toxins; or
- involves risks or exposure to accidents that young people may not recognize, 'owing to their insufficient attention to safety, or lack of experience or training'.

(Management of Health and Safety at Work: Approved Code of Practice 2000)

What you can do yourself

If you suffer from a sudden injury at work, or develop ill health from longer-term exposure to a hazard, you should consider taking action at work with the following steps:

- **Report** the incident to your workplace safety rep (where there is one) and line manager.

- **Record the incident/illness** in the accident book. If the health risk is serious, report the incident/illness either to a local authority or to an HSE inspector.
- **Contact your employer's** occupational health (OH) service if available.
- **Be better informed** – check your contract of employment, staff handbook or other policy documents.
- **Visit your GP**. Visit your local doctor and explain the full story (see Chapter 21). Explain why you think your health problem may be work-related.
- **Contact a community-based OH service** – if there is one near you (see Chapter 23).
- **Use a helpline**. Use your employer's employee assistance programme, if there is one. Contact a union helpline (see Chapter 25).
- **Keep a diary**. This may help you identify the source of the problem, and provide evidence if you decide to take legal action.
- **Use self-help guides**. Practical advice is available from voluntary bodies, trade unions and elsewhere – see our examples on safe use of your voice (page 99), back strain (page 71), bullying (page 164) and working safely alone (page 149).

And, if nothing happens, consider the following:

- **Take out a grievance**. If your line manager fails to resolve the problem, use your employer's internal grievance procedure.
- **Use your right to be accompanied**. You now have *the right to be accompanied by a union representative* when meeting your employer, irrespective of whether a union is formally 'recognized' by your employer – see Chapter 25.
- **Prepare your case**. Get together evidence of your condition, and how it is linked to your work. You could use a diary, evidence from your GP, or advice from your union representative.
- **Union support**. Many of the workers we have interviewed were supported by their union at crucial stages. This support includes:

- informal, confidential advice;
- representation at meetings with managers;
- guidance on specific rights at work; and
- access to union health and safety, and legal services.

Where an employer *recognizes* a trade union, safety representatives have a wide range of rights to carry out their functions (see box, page 38).

Taking legal action

If you have resigned from work, or just walked out, because of work-related ill health or injury, two courses of action are open to you. The first is claiming **unfair 'constructive dismissal'** under the Employment Rights Act 1996. Your claim must be submitted to an Employment Tribunal within three months of the day you last worked. But constructive dismissal can be very difficult to prove: be careful before you leave, always get some advice. The second course of action is claiming **personal injury compensation**, by making a civil claim in a County Court or the High Court (see page 259).

Breach of contract

Different legal tests apply in Employment Tribunal cases for unfair dismissal, and in the County or High Court in personal injury claims. But the *key test* is whether the employer's failure to take action involved a serious or fundamental *breach of contract*.

Every worker's contract of employment is made up of a number of terms, some written down, called 'express terms', some 'implied' and some brought into the contract because they are statutory entitlements, based on an Act of Parliament. Breach of contract could relate to any of the terms of the contract. **Express terms** are those you agreed with your employer directly, either in writing or orally. Express terms may include, for instance, an anti-harassment procedure. Some express terms may refer directly to statutory rights, eg on the safe use of display screen equipment. **Implied terms** are not directly agreed between you and your employer, but nonetheless they are *brought into* your contract. Some arrive by way of previous Court decisions, some by custom and practice in your workplace, and some by statutes. Implied terms are just as powerful as the 'express' terms, and include the *employer's duty of care* and the *duty of mutual trust and confidence* (see above, page 23).

Legal advice

Whichever course of action you decide on, you will first of all need legal advice. If you are a union member, consult your union rep or full-time officer. He or she may refer you to the union's specialist legal advice services. If you are not a union member, contact a law centre, citizens advice bureau (CAB) or solicitor (see Chapter 24).

Protection from victimization

Any employees who are *victimized* for taking up a health and safety issue at work, whether they do so on their own account or as recognized safety representatives, are protected under the Public Interest Disclosure Act (PIDA) 1998. The Act applies to all employees, regardless of their age or length of service, and to union and non-union members. It includes people who are victimized for following internal grievance procedures, as well as those making wider, public disclosures.

However, you need to 'reasonably believe' that the issue concerned is likely to endanger the health and safety of any individual worker, or group of workers. A victimized employee can complain within three months to an Employment Tribunal – see 'How to get safe, not sacked', *Hazards* magazine, **79**: www.hazards.org.

'Roving' safety reps

Unions, the TUC and the National Hazards Campaign have been campaigning for 'roving' safety reps to help address the relatively poor safety record in non-union workplaces. The TUC argues that non-union workplaces lack:

- the protection of a health and safety committee;
- employee consultation on safety;
- elected safety reps.

TUC commissioner on the HSC, Owen Tudor, suggests that a key factor is the steady decline of union recognition by UK employers over the past 20 years. Now, only a minority, perhaps one-third, of UK workers are consulted about their safety at work.

Tudor sees this as a matter of concern for both sides of industry, employers and workers, 'because employers agree with unions that involving the workforce is a better way of running the health and safety system. They are just not convinced that employers want or need all the other baggage that recognizing a trade union would bring.'

A way forward proposed by the HSC, with TUC support, involves union-appointed 'worker safety advisers', trained in health and safety. They would be able to inspect workplaces without a recognized union. Meanwhile, **individual** union members in an otherwise 'unrecognized' workplace now have the right to be accompanied at meetings with their employer concerning their health and safety at work (see Chapter 25).

Updating health and safety

Safety laws are constantly being updated. Whether through new evidence of ill health, public concern, new technology or European law. The development and improvement of health and safety policy is a key role of the Health and Safety Commission. Journals such as *Hazards, Labour Research* and *Occupational Health Review* provide regular updates and commentary on current developments.

As we go to press, the Government has announced plans to publish a draft corporate killing bill towards the end of 2003. In a critical reaction, the Centre for Corporate Accountability pointed out that this would be the third round of public consultations in 10 years on strengthening criminal law sanctions against negligent company directors. Just 15 company directors were convicted of health and safety offences between April 1999 and January 2003. Average fines were £2,656.

In 2002/03, the HSC published the Control of Substances Hazardous to Health (COSHH) Regulations that updated its controls on chemical and biological hazards; brought in new controls on asbestos; and introduced a new Approved Code of Practice on work-related asthma (see page 16). The Management of Health and Safety (Miscellaneous) Regulations 2002 made various changes to first aid, display screen equipment, manual handling and other Regulations.

The HSC is currently considering new initiatives on:

- **Smoking at work**. Three million people are exposed to the risks of passive smoking at work. Passive smoking may be causing the deaths of up to 1,200 people a year in the UK. In future, if the Government supports the HSC's recommendations, a new ACoP will protect workers with a health condition, eg asthma, and others who face the risk of tobacco smoke. It will cover bars and restaurants, and include control measures such as improved ventilation, as well as allowing complete or partial smoking bans at work. HSC advice is currently limited to advice to employers – see *Passive Smoking at Work.*

- **Consultation with safety representatives**. The HSC is considering new Regulations on the rights and facilities available to safety reps. Changes the TUC and affiliated unions want include a new duty on employers to consult with safety reps on risk assessments; and a new right for safety reps to have access to employees they represent, even when they don't work on the same premises.

The 'six-pack' of Regulations...

During 1989, six European Directives, collectively known as the 'six-pack', were issued by the European Commission. The following Regulations brought the Directives into effect in the UK:

- **The Management of Health and Safety at Work Regulations 1999**. The 'Management regs' first took effect in 1993. Main employers' duties include:

  - To make 'assessments of risk' to the health and safety of their employees, and to act upon risks they identify, so as to reduce them (Regulation 3). This duty on employers to carry out a risk assessment is likely to be highlighted by personal injury lawyers, to substantiate whether or not the employer has acted reasonably to provide a safe system of work.
  - To appoint competent persons to oversee workplace health and safety.
  - To provide employees with information and training on occupational health and safety.
  - To operate a written health and safety policy.

- **The Workplace Health, Safety and Welfare Regulations 1992**. The main provisions require employers to provide:
  - Adequate lighting, heating, ventilation and workspace, to be kept in a clean condition.
  - Staff facilities: toilets, washing, refreshment.
  - Safe passageways, eg preventing slipping and tripping hazards.
  - Adequate facilities and access for disabled employees.
  - Separate rest areas for smokers and non-smokers (Regulation 25).

- **The Display Screen Equipment Regulations 1992**. The main provisions apply to display screen equipment (DSE) 'users', defined as workers who 'habitually' use a computer as a significant part of their normal work. This includes depending on the use of DSE to do the job; using DSE for an hour or more continuously; and making daily use of DSE in this way. In 2002, the DSE Regs were expanded to cover TV editing technicians, security control room operatives, micro-electronics assembly operatives using DSE to view data, agency staff, and the use of laptops. Employers are required to:
  - Make a risk assessment of workstation use by DSE users, and reduce the risks identified.
  - Ensure DSE users take 'adequate breaks' (see also page 86).
  - Provide regular eyesight tests.
  - Provide health and safety information.
  - Provide adjustable furniture (desk, chair, etc).
  - Demonstrate that they have adequate procedures designed to reduce risks associated with DSE work, such as 'repetitive strain injury' (see page 61) and stress.
  - A *VDU Workstation Checklist* is included in HSE guidance.

- **The Personal Protective Equipment at Work Regulations 1992**. The main provisions require employers to:
  - Ensure that suitable personal protective equipment (PPE) is provided 'wherever there are risks to health and safety that cannot be adequately controlled in other ways'. The PPE must be 'suitable' for the risk in question, and include protective face masks and goggles, safety helmets, air filters, ear defenders, overalls and protective footwear.
  - Provide information, training and instruction on the use of this equipment.
- **The Manual Handling Operations Regulations 1992**. The main provisions require employers to:
  - So far as is reasonably practicable, avoid the need for employees to undertake any manual handling involving risk of injury.
  - Make assessments of manual handling risks, and try to reduce the risk of injury. The assessment should consider the task, the load and the individual's capability and take account of any increased risks to pregnant workers.
  - Provide workers with information on the weight of each load.

For further information, see page 67.

- **The Provision and Use of Work Equipment Regulations 1998**. The main provisions require employers to:
  - Ensure the safety and suitability of work equipment for the purpose for which it is provided.
  - Properly maintain the equipment, irrespective of how old it is.
  - Provide information, instruction and training on the use of equipment.
  - Protect employees from dangerous parts of machinery.

Reporting injuries and ill health

Employers are required to report a wide range of work-related incidents, injuries and diseases to the Health and Safety Executive (HSE), or to the nearest local authority environmental health department.

The Reporting of Injuries, Diseases and Dangerous Occurrences Regulations 1995 require an employer to record in an accident book the date and time of the incident, details of person(s) affected, nature of injury or condition, the person's occupation, the place where the event occurred and a brief note on what happened.

The following injuries or ill health must be reported:

- death of any person;
- major injury: including fractures, amputations, eye injury, injury from electric shock, acute illness requiring removal to hospital or immediate medical attention;
- 'over-three-day' injuries: involving someone off work for more than three days as a result of injury caused by an accident at work;
- specified diseases, including:
  - cramp of the hand or forearm due to repetitive movement;
  - beat knee, from physically demanding work;
  - carpal tunnel syndrome, involving hand-held vibrating tools;
  - hepatitis;
  - tuberculosis;
  - occupational dermatitis.

Rights of workplace safety representatives

The Safety Representatives and Safety Committees Regulations 1977 give recognized trade unions the right to appoint workplace safety reps. The Regulations require employers to set up a safety committee and to consult with recognized safety reps. Safety reps' rights include the right to:

- Take an active part in workplace risk assessments.
- Investigate potential hazards and 'dangerous occurrences', and examine the accident book.
- Investigate members' complaints.
- Carry out inspections of the workplace, at least every three months.
- Require their employer to set up and attend a safety committee.
- Be consulted on new working practices and new technology.

- Receive safety information from the employer, eg inspectors' reports, hygiene surveys, risk assessments.
- Attend union-approved training courses. Employers should provide access to a phone and office equipment, and paid time off work to carry out inspections, and to meet staff and other safety reps.

(TGWU Safety Representative's Handbook)

Claiming benefits

If you are suffering from a work-related injury or disease, you may be able to claim Industrial Injury Disablement Benefit, and other associated benefits.

Disablement Benefit is the main benefit payable as part of the industrial injuries scheme. The disability can be physical or mental. You can work and continue to receive benefit. It is payable weekly, can be paid in addition to wages and is not taxable. It doesn't matter how long you have worked for your employer. There is no time limit for making a claim, but the maximum back-dating is normally three months' arrears.

Be sure to report your accident or illness at work at the time it happens or soon after. This isn't essential to claim a benefit, but it makes doing so much simpler.

Claims for industrial injury and associated benefits are made to your local Jobcentre. For advice on claiming, contact your nearest CAB, law centre or union rep.

To qualify for benefits, you will be expected to have a Benefits Agency health assessment of your disability. The level of disability directly affects your benefit payment. You should seek expert advice.

For your nearest citizens advice bureau, see the phone book. For union contacts, see Chapter 25. You can download information on these benefits. Go to: www.dss.gov.uk/benefits/industrial.

Further information

Health and Safety Law 2003 (2003), Labour Research Department (tel: 020 7902 9813), price £4.50.

Most unions publish a safety rep's guide. A good example is the *TGWU Safety Rep's Handbook*, available from Transport House, 128 Theobald's Road, London WC1X 8TN (tel: 020 7611 2500).

To contact the nearest Health and Safety Executive office, ring the HSE Infoline (08701 545500). The following are available from HSE Books (tel: 01787 881165) or the HSE's Web site at www.hse.gov.uk:

- *Consulting Employees on Health and Safety: A guide to the law* (1996), free.
- *A Guide to Health and Safety Consultation with Employees Regulations* (1996), price £8.
- *Management of Health and Safety at Work: Approved Code of Practice* (2000), price £8.
- *Safety Representatives and Safety Committees* (ref L 87), price £5.75.
- *A Guide to the Personal Protective Equipment Regulations 1992.*
- *VDU Workstation Checklist.*
- *Passive Smoking.*

Your Rights at Work: A TUC guide, available from TUC, Congress House, Great Russell Street, London WC1B 3LS (tel: 020 7467 1294).

Sources

Hazards magazine, monthly, available from tel: 0114 267 8936 (Web site: www.hazards.org), price £15 a year.

Health and Safety Bulletin, monthly, available from tel: 020 7354 6761, price £12 (single copy).

3 Assessing risks at work

> Under constant pressure to finish tasks in allocated time, and to stick to times in work programme. Insufficient travel time between jobs. Domino effect of arriving late at clients' homes, or being delayed in their homes. Back injuries – emotional and time pressures often make us careless, and take unnecessary risks. The very fact that they are lonely/isolated increases the risk we put ourselves under – 'You're the only one who won't do that for me', is a common cry.
>
> (From a survey by a home carers' shop steward)

Any employer who is serious about preventing ill health and injury will regularly assess risk at work. To do so is an explicit requirement in the health and safety at work 'Management regs' (see page 35). *Five Steps to Risk Assessment*, one of the most important employer's guides published by the Health and Safety Executive, urges employers to seek out and eliminate hazards at work.

Risk assessment lies at the heart of the trade union approach to health and safety. UNISON's safety reps' handbook says:

> Risk assessment is a simple concept. It is the process of identifying what hazards exist in a workplace, and how likely these hazards are to cause harm, in order to decide what prevention or control measures are needed. If risk assessments are done correctly, they can mean that workers are properly informed about their working conditions, the risks involved, and how to avoid them.

In this chapter, we look at:

- employers' duty to assess risks;
- five steps to risk assessment;

- how to 'map' health risks at work;
- pregnancy and VDUs.

Employers' duty to assess risks

Despite clear legal duties on all employers to carry out risk assessments, HSE's safety studies show wide variations in compliance with the law. Large firms are more likely to carry out 'risk assessments' than small ones. Smaller organizations are noticeably less aware of the 'six-pack' Regulations (see page 35) than larger ones. Evidence that workplace injury and ill health is *twice as likely* in small and medium-sized firms is but one measure of the scale of the problem of risk *underassessment*.

Five steps to risk assessment

The HSE's guide to risk assessment says: 'If you are doing the assessment yourself, walk around your workplace and look afresh at what could reasonably be expected to cause harm. Ignore the trivial, concentrate on significant hazards which could result in serious harm.'

The HSE encourages employers to find out what their employees and union representatives think about health and safety because 'they may have noticed things which are not immediately obvious'.

Whoever is nominated by your employer to carry out the assessment must be a **'competent person'**. The Management of Health and Safety at Work Regulations define this as someone with sufficient 'knowledge and experience' to do the job properly. Recognized safety reps have the right to be consulted on the selection of a competent person.

The HSE advises employers to follow five steps when carrying out a workplace risk assessment.

Step 1: Identify hazards, ie anything that may cause harm

Employers have a duty to assess the health and safety risks faced by their employees. Where a union is recognized, employers must consult with safety reps as part of risk assessment. Nevertheless, employers retain ultimate responsibility for the assessment, and for any steps that they need to take to eliminate, or control, risks.

Your employer must be systematic in checking for possible physical, mental, chemical and biological hazards. UNISON's safety reps' guide groups hazards into four main categories:

- **hazards of the working environment**: the physical, mental, biological and chemical hazards we listed on page 29;
- **hazardous tasks**, eg cleaning, lifting, dealing with the public, driving;
- **locational hazards**: offices, libraries, depots, working off-site;
- **organizational hazards**: working hours, the system of work, staffing levels, employee participation.

Physical hazards include:

- 'Musculoskeletal disorders'. This is a general term for a wide range of work-related aches and strains. These include back pain from lifting, carrying or sitting awkwardly. And injury to the upper limbs from repetitive use of the hand, wrist and lower or upper arm – conditions known as repetitive strain injuries (RSI) or work-related upper limb disorders (WRULD).
- Injuries from physical attacks.
- Tripping and slipping hazards.

Mental (or 'psychosocial') hazards include:

- stress and overload;
- bullying;
- violence;
- lack of control over work; and
- long hours.

Step 2: Decide who may be harmed, and how

Identifying *who* is at risk starts with your organization's own full- and part-time employees. However, employers are also responsible for assessing risks faced by agency and contract staff, and members of the public on their premises.

Employers must review work routines in all the different settings and situations where their staff are employed. For example:

- Home care supervisors must take due account of their *client*'s personal safety in the home, and ensure safe working and lifting arrangements for their own *home care staff*.
- In a supermarket, hazards are found in the repetitive tasks at the *checkout*, in *lifting* loads, and in *slips and trips* from spillages and obstacles in the shop and storerooms. Staff face the risk of *violence* from customers and intruders, especially in the evenings (see page 133).
- In call centres, *workstation equipment* (desk, screen, keyboard, chair) must be suitably adjusted to suit each employee, including agency workers.
- Employers have special duties towards the health and safety of young workers, disabled employees, night- or shift-workers, and pregnant or breastfeeding women.

Safety mapping

Many unions now advise their safety reps to consider using a questionnaire survey for mapping health and safety 'hot spots'. The survey should cover employees' physical and mental well-being (see box, page 48).

Step 3: Assess the risks, and take action

This means employers must 'consider how likely it is that each hazard could cause harm. This will determine whether or not you need to do more to reduce the level of risk. Even after all precautions have been taken, some risk usually remains. What you have to decide for each remaining significant hazard is whether this remaining risk is high, medium or low, the HSE says.

Assessing risks with a points system

UNISON uses a simple four-point scale to 'score' the risks associated with any identified hazard:

- 0 = no risk;
- 1 = slight risk of not very serious injury;
- 2 = moderate risk: more people likely to be injured, or more serious injuries might occur;
- 3 = high risk: significant chance of serious injury.

Another system is the Risk Score for any hazard, found by multiplying the *likelihood* by the *consequences*. The matrix in Table 3.1 shows how the system works. Each hazard is scored against this matrix. It is drawn from real-life situations and can be applied to any hazard. The results are then discussed with management.

Preventing risks

The basic rules for employers who intend to tackle risks at work are set out in Regulation 4 of the Management of Health and Safety at Work Regulations. They introduce the idea of a 'hierarchy of control', following these rules:

- **Substitution**. First, if possible, employers should 'avoid the risk altogether', eg do the work in a different way.
- **Evaluation**. Risks that cannot be avoided must be evaluated through a risk assessment.
- **Prevention**. Any risks that are identified must be tackled at source, rather than taking the easy way out. If steps are slippery, they should be treated with a non-slip surface, or replaced. Putting up a warning sign is not being serious.
- **Adapting work**. Adapt the work to the worker. Employees should be consulted if work practices are to change. 'Aim to increase the control individuals have over their work', the Regulations say. Ensure office chairs and display screen equipment are adjustable to the individual. Plan work involving a computer to include regular breaks. For monotonous or routine work, introduce work variety and greater control over work.
- **Training and information**. Provide training in safe working systems, and information on likely hazards and how to avoid them.

Table 3.1 Risk scoring system
Values in each box are from multiplying Likelihood by Risk

	Likelihood Category: Score 1 to 5				
Risk Category Score 1 to 5	**1: almost impossible**	**2: unlikely**	**3: possible**	**4: likely**	**5: almost certain**
1: no injury	1	2	3	4	5
2: first aid needed	2	4	6	8	10
3: off work for less than three days	3	6	9	12	15
4: off work for over three days	4	8	12	16	20
5: major injury	5	10	15	20	25

(Source: UNISON)

- **Welfare**. Provide social and welfare facilities, eg washing facilities for the removal of contamination, a rest room.
- **Protective equipment**. If elimination and substitution do not succeed in controlling risks, then employers must issue personal protective equipment, eg appropriate eye protection or gloves.

Step 4: Make a record of the findings

Employers with five or more staff are required to *record in writing* the main findings of the risk assessment. This record should include:

- details of any hazards noted in the risk assessment;
- action taken to reduce or eliminate risk levels;

The record provides proof that the assessment was carried out, and the basis for a later review of working practices. The risk assessment is a working document. It should not be locked away in a cupboard.

Step 5: Review the risk assessment

Reviewing a workplace risk assessment is important for three main reasons:

- To ensure that agreed safe working practices continue to be applied, eg that management's safety instructions are respected by supervisors and line managers.
- To take account of any new working practices, or new machinery, eg new, more demanding work targets may have prompted higher absence rates, highlighting the need for a new risk assessment.
- To take account of staff turnover: high staff turnover places strain on agreed safety procedures. A major finance sector call centre, with 720 staff, lost nearly 80 per cent of its workforce in one year. The 550 people taken on during the year needed training in agreed safety routines.

Sickness absence records and the accident book should be reviewed regularly, to help identify unexpected health and safety 'hot spots', where a new risk assessment may be required.

How to 'map' health risks at work

Questionnaire surveys help safety reps to 'map' *collective* health risks in the workplace. The example in Figure 3.1 lists some typical job-related health conditions, and asks how often they are felt. You can adapt the survey form to your own needs. *Body-mapping* is a further technique used to identify occupational health issues (page 74). Both methods are useful in helping safety reps to identify priorities for action in discussions with management.

Questionnaire Survey Form

1. Do you suffer from any of the following? (please tick as appropriate)

	Never	Hardly Ever	Sometimes	Often
Headaches				
Eyestrain				
Tension				
Tiredness				
Aches/pains in hands				
Aches/pains in wrist				
Backache				
Rashes or itching				
Difficulty sleeping				

2. At work, do you feel (please tick as appropriate)

Irritated				
Angry				
Frustrated				
Helpless				
Anxious				
Depressed				
Confused				
Bored				
Happy				
Enthusiastic				

3. Do you think these symptoms are related to your work or your workplace?

---

4. Have you any other illnesses that you think may be related to your work? Give a brief description; include whether you have visited your GP to discuss the issue.

The questionnaire can be adapted to suit many situations. For example, in a call centre, questions might be added on ringing in the ear (tinnitus), or voice loss. Home care workers might need questions on safety arrangements when working out of the office, or incidents of violence whilst on duty.

(Source: UNISON)

Figure 3.1 Questionnaire survey form

Pregnancy and VDU risks

At present, there is no scientific evidence to suggest a link between miscarriages or birth defects and working with computers (also known as a visual display unit: VDU). However, to avoid problems caused by stress and anxiety, an employee who is pregnant or is planning to have children should approach their supervisor, team leader or safety adviser to discuss their concerns.

Nottingham City Council (quoted above) uses a 'risk assessment' approach for pregnant staff concerned about the health risks of using a VDU:

- There is no clear evidence of health risk, it suggests.
- But, to avoid stress and anxiety:
  - Approach your team leader or supervisor, to discuss your concerns.
  - If your concerns remain, a transfer to other non-VDU work should be made available during pregnancy.
  - Transfer on your existing grade (salary and conditions), with no demotion.
  - But, your temporary transfer may be to another section or department.

(Nottingham City Council Health and Safety Advisers' Unit, tel: 0115 915 6747)

Further information

Amicus – AEEU Guide to Risk Assessment, available from Amicus (tel: 020 8462 7755).

Five Steps to Risk Assessment, Includes A–Z list of the 20 HSE local offices, with phone numbers and addresses, free. HSE guides, available from HSE Books (tel: 01787 881165).

Making Your Workplace Safer: GMB safety representatives' guide and *Guide to Risk Assessment,* available from GMB (tel: 020 8947 3131).

Risk Assessment: A guide for UNISON safety representatives includes: a summary of risk assessment Regulations; practical examples of risk assessment techniques; and a do-it-yourself health survey kit. Available from UNISON health and safety unit (tel: 020 7551 1446).

Part 2

A pain in the workplace

4 A job at the supermarket

As Gita takes the single step down to her kitchen, subconsciously she places a hand on her right hip for support. Everyday tasks such as washing up – arms held level, wrists and hands rotating – are too painful to sustain for more than 10 minutes at a time. She has dropped plates, cups, a container of cooking oil on the kitchen floor. She cannot cook, clean, do the garden, use a vacuum cleaner, lift, carry shopping or dress without pain.

The Health and Safety Executive defines her condition as a 'musculo-skeletal disorder', that is, a disease affecting soft tissues and joints. Without early diagnosis and intervention, this condition can seriously damage the quality of life of the sufferer, not only at work, where the condition may originate, but in many everyday activities.

Gita worked as a supermarket checkout operator for 15 years. After 10 years in the job, in 1995, the first painful episode kicked in. Over the next two years she had a succession of appointments and meetings with her GP, a consultant rheumatologist, two physiotherapists and her employer. But not until she met her second physiotherapist did any health professional advise her that her condition might be linked to her job.

An occupational health specialist says she is suffering from 'work-related tenosynovitis' of her right wrist and right elbow joint ('teno' means tendon, and 'itis' inflammation). She also has 'cervical pain syndrome' ('cervix' means neck, and 'syndrome' a pattern or typical combination). It affects the whole of the upper right side of her body: hand, wrist, lower arm, elbow, upper arm, right shoulder, neck and back.

Gita's story

Gita's struggle to come to terms with her injuries begins and ends with her employer. Mistaken diagnosis, and delays in NHS appointments and treatment played a part. If there is now a conclusion of sorts, it is in large part due to her union's support.

Gita has good days and bad days.

> If you don't take painkillers, and don't do anything but just sit, then you can be all right. I do try to cut down on things, like I won't cook a proper meal every day. This week, Monday, I cleaned the house a little. But on Tuesday I went shopping, and, after just lifting a 2.5-kilogram bag of potatoes off the shelf, I couldn't sleep at night, it was that bad.

Gita has also developed asthma, an allergic reaction to the painkillers and medication prescribed by her doctor.

A safe system of work?

Gita started full time as a checkout operator at a London branch of a leading supermarket in October 1985. At that time, her main work target was to scan at least 1,200 items an hour, or 20 items a minute. A computer monitored the hourly performance of each employee.

> I must have handled tons of goods every week. You are lifting heavy items, scanning them, passing them through the checkout. You can have cat litter at 10 kilos, a 7.5-kilo bag of potatoes, 12 litres of water in a pack.

She always sat on the same side of the till, facing customers the same way. 'Everything came from my right hand to my left hand, always right to left. I have no pain or anything wrong with my left side.' The weighing scales were also situated to her right. 'You twisted round to weigh things, including really heavy items like potatoes. And you bent and twisted to pick up carrier bags.'

NHS mistakes and delays

She felt the first sharp pains in her right hand in August 1995. She visited her GP, who prescribed a course of analgesic tablets, and the pain subsided.

'Then, in June 1996, I suddenly started dropping things at the till, like customers' credit cards. Because the pain came suddenly, it made me jump.' She describes the pathway of the pain as shooting from her right wrist to her elbow. 'It was just like somebody stabbed you. I couldn't control it.'

Her GP referred her to hospital for X-rays, and she returned to work at the tills. A fortnight later, her doctor reported that the X-rays revealed no signs of fracture, arthritis or rheumatism. She did not connect her condition with her job, but unfortunately nor did her doctor. 'He asked me where I worked, and I told him. But he didn't ask me what work I did.'

Gita waited six months for a consultant's appointment. Meanwhile, she didn't miss a day's work. Once, when the pain was particularly uncomfortable, she informed her supervisor, who suggested that it was caused by arthritis. 'But even I didn't know what it was. I said that my X-rays showed I haven't got arthritis. I just asked them to get me off the till, just for a short break. Literally, I would have to beg them sometimes.' If she had a long weekend off, the pain would subside by the time she was due back at work. This is how she coped.

Her first appointment with a consultant rheumatologist was in January 1997. The consultant prescribed a wrist support and physiotherapy. When the store manager noticed the wrist support, he said they needed to meet, but before they could do so he was transferred to another store.

Her first physiotherapy session was in March 1997, nine months after she had dropped the credit cards at the till. 'I told her about my job, but she didn't say anything in connection with my work. It seemed to me that she didn't really want to know.'

One day, needing time off to attend a hospital appointment, she went to ask the store manager how she should make up the hours. The store's shop steward, Zoe, who happened to be in his office, overheard his reply and reminded the manager that full-time staff did not have to make up time for hospital appointments. Zoe had stood up for her. The manager relented, and Gita decided to join the union.

Gita was returning straight from physiotherapy to work on the tills. She asked her physiotherapist whether this wasn't cancelling the benefits of the treatment. 'I asked if it was possible that she could write a letter to my manager. I had already talked to him, but they wouldn't let me off the checkout.' The physiotherapist agreed to ask the consultant about this. It is unclear why her consultant had not already suggested this option to Gita.

What's repetitive strain syndrome?

> But the next time I went I had a new physiotherapist. He was very good. He went straight to my neck, checked me through and sent me for another X-ray. He said that I was suffering from 'repetitive strain syndrome'. I said, 'What's that?' I wrote it down. That's the first time anyone told me. I asked again about the letter. He advised me to talk to the checkout supervisor, to ask for breaks from time to time. I told him that I had already done that. I took the letter with me to work.

She handed the consultant's letter to the personnel manager, but no action was taken until a meeting a few weeks later with the company's occupational health (OH) adviser. This took place a whole year after Gita had dropped the credit card. The OH adviser remarked to the personnel manager that it didn't make sense for Gita to go for physiotherapy and then return to work on the tills.

The company agreed to transfer Gita to lighter, alternative duties on the cigarette kiosk. She fully expected to take up her new position when she returned from holiday. Nevertheless, on her first day back her supervisor directed her to work on the tills, because there was nowhere else to put the staff on the kiosk.

Gita was not prepared to return to the checkout against occupational health advice. She started to help customers with packing, and went to see the personnel manager as soon as she arrived that day. The personnel manager said that the grocery manager needed a code checker. This was a surprise, as this option had not been discussed with the OH adviser, and Gita was not trained for that work.

> I did code checking for a while, but as soon as it came to lifting and checking the larger tins, twisting and turning them to check the bar codes, the pain came back within 5 or 10 minutes. I just burst into tears. I went upstairs and took a couple of painkillers. Instead of putting me on the kiosk, they put me there.

She met the grocery manager with Zoe, the union rep, who asked why she was lifting tins, that it wasn't a light job, as agreed.

> That evening, I went to my doctor. As soon as I got there I started crying, because the pain was so bad in my neck. He gave me a week's sick certificate. That was the first week I had ever had off. It was not only to do with my wrist, but also my neck. I cannot describe the pain. The first time I cried with the pain, actually, was that day.

Over the next six months, she worked on store security, in the canteen, the delicatessen counter, the grocery, and even the tills – almost everything except the agreed post on the kiosk. Another idea, office work, failed when some other staff accused her of queue jumping, as they wanted an office job.

One morning, in early June 1998, around the second anniversary of the credit card incident, Gita recalls that she woke up early with a severe neck pain. It was so bad that she could neither move her neck nor go back to sleep. But, hoping it would go away after taking painkillers, she continued to go into work for the following nine days. The analgesic was ineffective, despite a long weekend's rest.

The following Tuesday, her doctor prescribed a new, stronger painkiller, which seriously upset her stomach. 'A week later, I went back to my GP, because something was really wrong. He gave me another sick note.' She was then unable to work for six months. During this time, she fell into the catch-22 of benefits and fitness for work.

Her continued entitlement to Incapacity Benefit depended on her passing the then fit-for-all-work test, carried out by a DSS-appointed doctor. She failed this test. The doctor decided that she was not unfit for *all types* of work. So her Incapacity Benefit was stopped, in March 1999, and she became dependent on her disabled husband's income (about £60 a week) and his Disability Living Allowance. The family income was almost entirely consumed by mortgage and insurance payments. In October 1999, her own DLA was stopped.

Return-to-work plan fails

After Zoe, the union rep, moved to another store, a full-time official from the TGWU took over Gita's case, arranging meetings with OH advisers and the store management. Finally, in April 1999, a return-to-work plan was agreed, involving a phased increase in her working hours. The plan started with three hours on the first day, followed by a day's break, and a further three hours at work the next day: six hours' employment a week in total. Her role would be checkout supervisor, but she was to be paid the standard checkout operator's rate of £5 an hour, or £30 a week.

But the new arrangement was not a success. Gita was not given the full range of supervisor's duties, eg she did not have a set of till keys to deal with billing errors. Some supervisors made her feel unwelcome; one instructed her to start loading customers' trolleys.

The company tried, without further consultation, to end her employment. This time, the union officer persuaded the employer to follow its own disability awareness guidelines. They agreed to seek advice from a Disability Employment Adviser from the Employment Service, on other possible options. The Disability Employment Adviser offered Gita help with training in the use of voice-activated computers, possibly leading to work on the organization's intranet shopping service. She undertook an initial two-day training course, but no job was offered at the end of it.

The union suggested other positions, such as coffee shop cashier. But the company refused because cashiers also had to clear tables and serve food.

Gita says: 'But you can't tell unless you try, and they never really let me try out another job, not the kiosk, not the canteen, not the office job, not the computer job.'

From the union's perspective, it appeared that 'for every idea we discussed, they seemed to produce a reason why it wasn't feasible. That was backed up by their OH advisers.' Altogether, Gita saw three company OH specialists. 'The first one was good, very helpful,' she says, 'but by the time I saw the third one, she wasn't interested. She didn't want to answer my questions, or to listen to me. She was more interested just to see me go, and that's it.'

At a final meeting, management restated its intention to terminate Gita's employment, and offered her a medical severance package, similar to the company's redundancy terms. Taking into account her length of service and outstanding holiday entitlement, she was offered an £8,500 severance payment, roughly one year's full pay.

Gita refused the offer and, supported by union solicitors, submitted a personal injury claim against the company for negligence. A specialist medical opinion, submitted as evidence in the County Court action against her employer for personal injury, concludes that, 'on the balance of probability, her condition is related to her work, especially at the checkout till... Repetitive lifting of heavy items caused joint and neck strain... her symptoms have become chronic, and are very likely to continue in the long-term.'

In November 2000, she accepted an out-of-court settlement of over £36,000, with her former employer also refunding to the DSS the benefit she had received over the previous 18 months. The settlement appears to be on the high side among the many hundreds of similar cases for checkout operators settled out of court.

Gita is now attending a pain management clinic, and expects to be referred for further treatment. She says, 'There are 20 working years left until I retire. Anything I do even now hurts. What kind of life is this?'

5 Aches and strains: the danger signs

Currently, over 1.1 million people in Britain are suffering from work-related injuries to muscles, tendons and nerves. These disorders affect the hands, wrists, arms, shoulders, lower back and legs. Soft tissues connecting muscles to bone (particularly the tendons), muscles themselves and the associated nerve supply are affected.

Taken together, these injuries are called musculo-skeletal disorders (MSDs). Some conditions are known by the more familiar name of repetitive strain injuries (RSI). Injuries to the upper body are also called work-related upper limb disorders (WRULDs).

The HSE estimates that, each year, around 12 million working days are lost through musculo-skeletal disorders caused, or made worse, by work. Employees affected take an average of 19 working days off sick each year. The back is the most vulnerable area, followed by the upper limbs and neck.

In this chapter and the next, we examine the symptoms and causes of MSDs suffered by manual workers, particularly supermarket checkout operators. In Chapters 7 and 8 we focus on RSI and other health risks for office and call centre workers.

Here, we look at:

- musculo-skeletal disorders;
- repetitive strain injury;
- preventing injury: the law;
- women and manual handling injuries;
- a healthy back; and
- reporting pain and taking action.

Musculo-skeletal disorders

The symptoms of musculo-skeletal disorders depend on the exact condition, as Table 5.1 shows, but usually include:

- aches and sharp pains in the hands, wrists, arms, shoulders and back;
- restriction of joint movements;
- loss of grip as a result of pain; and
- nodules and swellings at the site of the injury.

In the early stages, there may be no visible sign of bruising or swollen joints. The onset of symptoms is often gradual, so that one response is for sufferers to adapt the way they work to reduce the pain. They may pass the problem on to another finger, hand, wrist or arm, or adopt another equally unsafe way of working. From such beginnings, a more complex set of problems can set in.

MSDs: how to recognize them

Table 5.1 and Figures 5.1 and 5.2 are taken from the GMB booklet, *Don't Take the Strain: A GMB guide to work-related upper limb disorders*. Table 5.1 identifies the main conditions, their symptoms and causes.

Trade union concerns over the suffering caused by repetitive strain injury were first highlighted by the GMB general union in *Tackling teno*, published in 1985. One type of RSI is tenosynovitis (or 'teno' for short), the inflammation of tendon sheaths in the hands, wrists or arms caused by repetitive movements. The GMB's report made a major contribution to increasing awareness and understanding of RSI. The wider issues of back strain and musculo-skeletal disorders are now a major priority for the TUC.

Repetitive strain injury

Injuries to the hands, wrists and upper limbs can develop into a condition known as 'diffuse RSI'. There may no longer be a

Table 5.1 Recognizing musculo-skeletal disorders

Condition	Body Zone	Symptoms	Typical Causes
Bursitis Inflammation of the soft pad of tissue between skin and bone, or between bone and tendon. Called 'beat knee', 'beat elbow' or 'frozen shoulders' at these locations.	Knee elbow or shoulder	Pain and swelling at the site of the injury	Kneeling, pressure at the elbow, repetitive shoulder movements
Carpal tunnel syndrome Fluid and tissue pressing on the nerves that pass up the 'carpal tunnel at the wrist.	Wrist	Tingling, pain and numbness in the thumb and fingers, especially at night and weakness in the hand	Repetitive action with bent wrist. Use of vibrating tools
Dupuytren's contracture Thickening of tissue under palm of hand, causing fingers to curl up.	Hand	Occasional burning pain, gradual development of nodules on palm. Gradually, extension of fourth and fifth fingers becomes impossible	Can be hereditary condition, or caused by manual handling
Epicondylitis Inflammation of the area where bone and tendon are joined. Called 'tennis elbow' when it occurs at elbow.	Elbow and other locations	Pain and swelling at the site of the injury	Repetitive work, often in strenuous jobs like joinery, plastering and bricklaying
Ganglion A cyst at a joint or in a tendon sheath. Usually on back of hand or wrist. Often accompanies RSI.	Hand wrist	Hard, round, small swelling, usually painless	Repetitive work, frequent movement
Osteoarthritis Damage to joints resulting in scarring at the joint and growth of excess bone.	Spine, neck and related joints	Stiffness and aching	Long-term overloading of spine and other joints

Table 5.1 *continued*

Condition	Body Zone	Symptoms	Typical Causes
Tendonitis Inflammation of the tendons themselves, which can also lead to the tendons 'locking' in the sheaths, so that fingers, hands or arms cannot move easily.	Hand, wrist, arm	Pain, swelling, tenderness and redness of hand, wrist and / or forearm. Difficulty in using hand	Repetitive movements
Tension, neck or shoulders Inflammation of the neck and shoulder muscles and tendons.	Neck and shoulder	Localized pain	Having to maintain a rigid posture
Trigger finger Inflammation of tendons and / or tendon sheaths of the fingers, causing the finger to lock before jerking into position.	Finger	Inability to move fingers smoothly, with or without pain	Repetitive movements, coupled with having to grip too tightly for too long, or too frequently

(Source: adapted from *Don't Take the Strain: A GMB guide to work-related upper limb disorders*)

specific 'location' of injury. Unlike recognized disorders, such as tenosynovitis, there are no objective clinical symptoms that allow a doctor to diagnose a specific condition. Nevertheless, a recent court case found in favour of five women suffering from acute upper limb disorders (ULDs). The judge in the case recognized 'diffuse RSI' as a distinct type of injury, even though accepting that diagnosis presented particular difficulties. This judgment does not set a precedent, in that it was limited to the facts of the case itself. However, it will encourage others suffering from similar conditions to take action.

All in the mind? The judge didn't agree

In 1999, the Court of Appeal upheld the claim by five Midland Bank workers that they developed work-related upper limb

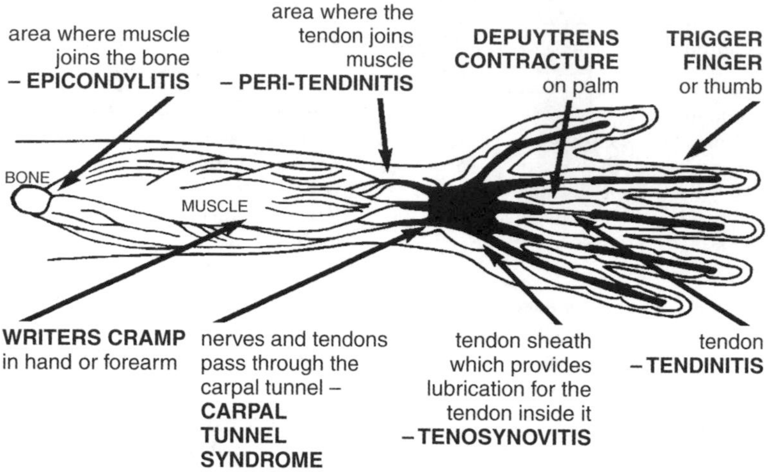

Figure 5.1 Hand and arm conditions
(Source: *GMB*)

disorders caused by repetitive work under intensive pressure. The five women worked as data encoders at the bank's Frimley service centre, processing the data from thousands of cheques every day. Evidence presented in court showed that they were given insufficient work breaks; the employer failed to provide footrests; and the women adopted poor arm postures in carrying out their tasks. They encoded thousands of data items an hour.

The medical evidence showed that they were suffering from 'diffuse' RSI, in that their work-related upper limb disorders did not present as clear physical conditions. The women suffered tissue damage to the muscles and nerves within the affected arm due to the 'prolonged static posture' they had to adopt whilst encoding data. The judge rejected the bank's assertion that the pain experienced by the five women was psychosomatic in origin, ie 'all in the mind'. Unifi, their trade union, successfully represented nine other staff in the wake of this decision (Case summary, Court of Appeal backs diffuse ULD claimants, *Health and Safety Bulletin*, **282**, October 1999).

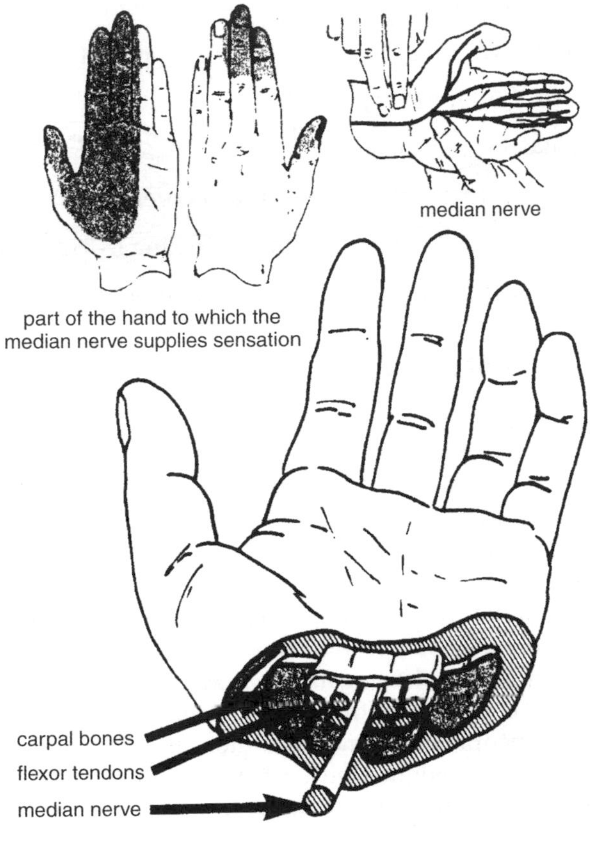

Figure 5.2 Hand conditions
(Source: *GMB*)

What causes RSI?

The five main risk factors associated with RSI are:

- **repetition** of the same sequence of movements many times an hour or day;
- **speed**: having to work very fast;
- **force**: having to use appreciable force;

- **lack of control** over the order of tasks or working methods;
- **postures**: working in awkward or tiring positions.

(*Hazards*, **73**, January 2001)

Working like this can cause harm:

- using hand-held, vibrating tools eg powered tools;
- working in cold environments, eg packing frozen items;
- inadequate training, leading to unsafe practices.

Young people at high risk

According to *Hazards* magazine, young workers are more at risk on all five of the above 'risk factors' than any other age group. In 'When work is a pain', the magazine analyses HSE injury data to show, for example, that young workers aged between 16 and 24 years were much more likely to be carrying out repetitive work at high speeds.

Nearly 4 million young people aged between 16 and 24 are in work, yet young workers often don't know their rights, are less confident about complaining and may well not be in a union. The Government emphasizes the use of computers in schools, and many young people use home computers and TV games. But they receive no safety guidance, either from school or manufacturers. They enter the world of work with their eyes closed to the risks of computer, keyboard and mouse. TUC is seeking to persuade leading computer retailers to issue a joint leaflet on display screen safety.

Preventing injury: the law

Each of the 'six-pack' of Regulations (see page 35) includes duties on employers that, directly or indirectly, provide protection for workers at risk from musculo-skeletal disorders. The common thread running through these Regulations is the need for a proper risk assessment.

The Manual Handling Operations Regulations 1992

The Regulations set out a **hierarchy of control measures** that employers should work through in order to prevent, or reduce, the risk of injury to their employees from manual handling of loads. The stages are:

- **Step 1**: **Avoid** the need for any manual handling involving risk of injury, 'so far as is reasonably practicable'. This may include mechanization, redesigning the tasks or breaking down the load.
- **Step 2**: **Assess the risks:** Where manual handling tasks cannot be avoided, employers must review the *risk factors associated with manual handling*:
  - the task;
  - the load;
  - the working environment; and
  - the individual's capability.
- **Step 3**: **Reduce the risk of injury**. After the risk assessment, introduce safe systems that minimize risks. The Regulations do not specify a maximum weight to be lifted. But employers must take steps to reduce manual handling to its lowest practicable level. They must provide employees with information on the weight of each load, and the heaviest side of any load.

A guidebook from the shopworkers' union, USDAW, shows the sort of weights that are likely to cause injury (illustrated in Figures 5.3 and 5.4). The weights are not meant to be interpreted as 'safe limits'. People may still be injured lifting lighter loads if other 'risk factors' are present, eg an awkward lifting position, or individual capability. When employees are handling the kinds of weights shown in the figures, then a risk assessment is likely to be needed. The figures assume that the worker is lifting easily held, compact loads in ideal conditions.

Women and manual handling injuries

It is often wrongly assumed that jobs that are traditionally done by women are 'lighter'. In fact, women are at as much risk as men from manual handling injuries.

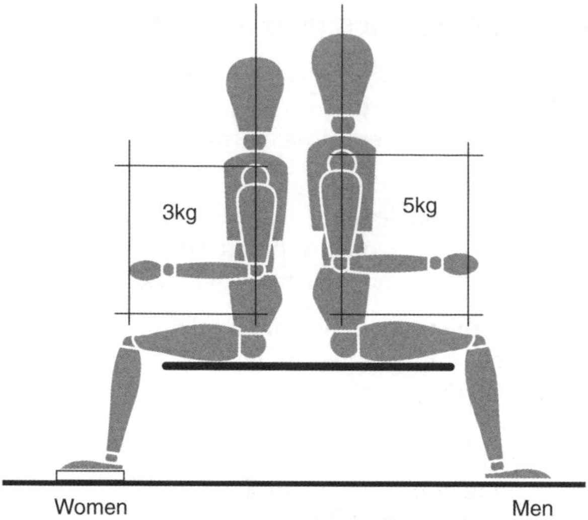

Figure 5.3 Handling while seated
(Source: *USDAW*)

True, as the GMB union says in *Working Well Together: Health and safety for women*: 'Women employees should not be asked to lift the same weights as their male counterparts without a specific assessment of the work involved. As a rule of thumb, women workers should only be expected to lift two-thirds the weights of their male counterparts.'

However, the GMB says there is such a 'wide overlap' between the sexes that there is no point in trying to identify tasks that 'only men' should do. If a load is too heavy for a fit woman to handle, it will be equally dangerous for a large number of men.

Some jobs require particular fitness, height or strength. But, in general, employers should aim to *fit the work to the worker*, and not the reverse. See also the HSE's Web site for advice on manual handling risks for new and expectant mothers: www.hse.gov.uk/mothers.

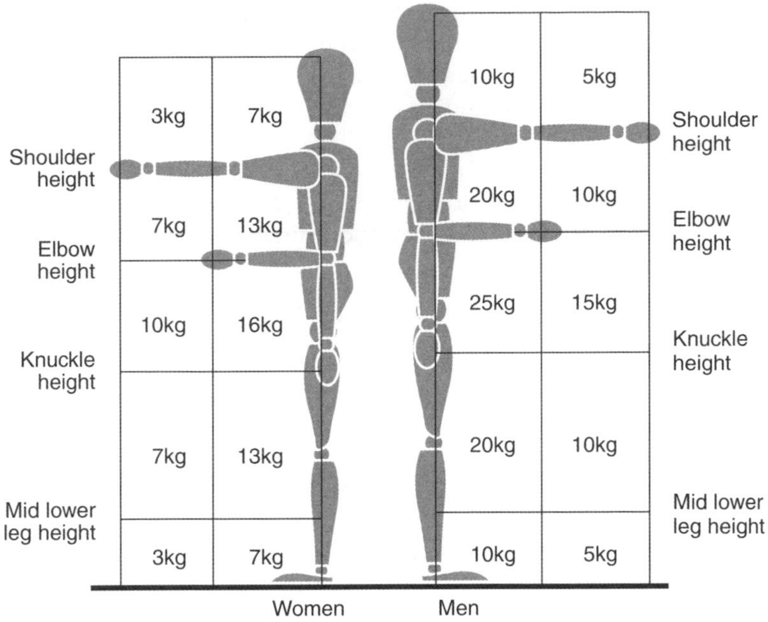

Figure 5.4 Lifting and lowering
(Source: *USDAW*)

A healthy back

A Government backcare initiative, called 'Back in work', is aiming to spread good practice by funding joint union/management projects across key industries with a high incidence of back injuries, eg textiles. It is also encouraging large employers to share their occupational health services with smaller organizations in their area, the so-called 'good neighbour schemes' (see *Employee Health Bulletin* 15).

Self-help techniques

The most likely sources of back pain are from damage to muscles and nerves (leading to a condition called 'spondylitis') and injury to discs (the shock absorbers in between each vertebra). According to registered osteopath, Garry Trainer, some of the most hazardous jobs, from a back care point of view, are those that involve prolonged periods of sitting. In *The No-Nonsense Guide to a Healthy Back*, he covers these issues:

Sitting posture

- Maintain good posture while sitting at work. Try to sit upright as much as possible, with abdominal muscles pulled in for support.
- Check that both feet are flat on the floor, or on a suitable footrest.
- Don't cross your legs – this restricts circulation and puts your spine out of alignment.
- Keep forearms in a horizontal position, and wrists supported when typing.
- At your desk, position the phone so that it is on the side that you answer it. Avoid crooking the phone in between your head and neck. If you use the phone a lot, a headset is a good idea.
- The more you sit the more you need the counterbalance of physical activity outside work, eg a brisk walk at lunch break, swimming after work.

Lifting posture

- Keep your back straight.
- Allow your thigh muscles (quadriceps) to do the work, rather than your back.
- Before lifting an object from the floor, take a wide stable stance and squat down. Think before you lift.
- Keep your head up, back straight and stomach pulled in.
- Grasp the object firmly; pull it close to your body.
- Stand up in one slow, smooth movement.
- Keep your head up; push with your legs.

- When carrying a heavy object to another location, make sure you have prepared the route by looking out for obstacles.

Driving posture

Driving is one of the most common places for back pain to start. Added risk factors include shoulder and back tensions after driving at night, in heavy traffic or for prolonged periods.

- Concentrate on good posture. Keep the steering wheel quite close, as this will create less strain when you turn the wheel.
- Use a lumbar (lower back) support in the seat. A neck rest is also a good idea.
- Take care getting in and out of the car. The safest movement is to sit first and slowly swivel round. When leaving, put both feet on the pavement before standing.
- After a long car journey, go for a long walk. This is not the time to slump in front of the TV!

Stretching

Stretching is particularly beneficial for your back. It keeps it mobile and reduces most muscular tension. Aim to stretch every day, so that it becomes a habit, like brushing your teeth.

If you suffer back strain at work

Some aches and pains in muscles and joints can just mean that muscles are tired. But pain and discomfort that doesn't go away is likely to be related to your work. It may be time to act:

- **Report** your symptoms to your employer.
- **Contact an occupational health adviser** – your employer may provide this service, or there may be one in your area (see page 252).
- **Contact** your union rep, if available.
- **Record any incident** in the accident book.

- **Visit your GP**. Tell your GP about the type of work you do. Check our advice on consulting your GP, see page 228).
- **Rest or exercise?** Your GP may advise a short period of rest, preferably with an ice pack. But long periods of rest can worsen the injury by weakening affected muscles or tendons. Painkillers prescribed by your GP may be necessary but, as the GMB guide points out, 'hiding the pain can prevent the body's natural alarm bells from warning you about potentially irreparable damage'. For persistent pain, the GP is likely to refer you to a specialist. If so, you may also need a doctor's letter to your employer, asking for alternative work on lighter duties.
- **Returning to work**. If you are off sick, even for a short period, with a work-related illness or injury, you may not be able to return to your original job straight away. The Disability Discrimination Act 1996 established new duties on employers to make 'reasonable adjustments' for people with disabilities (see Chapter 20).
- **Be accompanied at meetings with management:** for details of your rights, see page 271.

Taking legal advice or action

If the pain or discomfort does not ease after a rest from work, treatment or medication, you may need to seek legal advice about making a claim for personal injury (see Chapter 24).

Further information

Back Care Awareness, Newcastle City Council Back-in-work initiative, available from Occupational Health Unit (tel: 0191 211 5215).

Do You Suffer from Aches and Pains in Silence?, available from TGWU regional offices.

Ending Back Pain from Lifting. Individual copies are available from the Health and Safety Unit, UNISON, 1 Mabledon Place, London WC1H 9AJ (tel: 020 7388 2366).

Manual Handling: GMB safety reps guide and *Home Care Staff: Health and safety* and *Working Well Together: Health and safety for women*, GMB Publications, Health and safety unit (tel: 020 8947 3131; Web site: gmb.org.uk) A4 leaflets, free; larger booklets, price on request.

Preventing Manual Handling Injuries, USDAW (tel: 0161 224 2804; Web site: www.usdaw.org.uk).
Understanding Back Pain, Jayson, Professor Malcolm, in the British Medical Association *Family Doctor* series, available from pharmacists, supermarkets or direct from Family Doctor Publications (tel: 01295 276627), price £3.50.
Watch Your Back: A guide to preventing back pain and injury at work, available from Labour Research Department (tel: 020 7902 9813), price £3 (£10 commercial bodies).

Sources

The No-Nonsense Guide to a Healthy Back, Trainer, Garry and Alexander, Tania, available from Fingertips Press (tel: 020 7224 1750), price £4.99.
Preventing back problems in the textile industry, 2000, *Employee Health Bulletin*, **15**, available from tel: 020 7354 6761, price £12 (single copy).
Putting Back Strain on the Map is available from TUC (tel: 020 7467 1248).
RSI: A trade unionists' guide, Labour Research Department (tel: 020 7928 3649; Web site: www.lrd.org.uk), price £3.15.
When work is a pain, *Hazards*, **73**, available from tel: 0114 276 5695, price £10 a year. Go to: www.hazards.org.uk.

6 Body mapping

When the supermarket introduced a new checkout design, most staff began to report increased aches and pains in their neck and shoulders. With the agreement of the personnel manager, the store's union, USDAW, carried out a staff questionnaire survey, and found that almost every employee was suffering, especially shorter workers because of the extra stretching involved at the new tills. The findings led to joint consultations with management on a redesign of the checkouts.

'Mapping' the risks

A risk assessment is the starting point for identifying and tackling any work-related health issue (see page 41). New 'mapping' techniques, involving the active participation of workers, are helping to raise awareness of occupational health issues. Mapping is now an essential tool in risk assessment. It helps workers and management alike to identify the early signs of health problems before they cause widespread illness and sickness absence.

Using the practical example of manual handling hazards for supermarket checkout workers, this chapter looks at:

- body mapping;
- workplace mapping, using employee questionnaire surveys;
- hazard mapping;
- working safely in a supermarket;
- HSE's advice on safe working;
- self-help advice at the checkout.

Body mapping

The TUC, *Hazards* magazine and national charity BackCare promote body mapping techniques to encourage workers to locate their aches and pains on body charts. The technique was developed by unions in North America, and was promoted in the UK by *Hazards* magazine. As a group exercise, carried out by employees in the same kind of occupation, it encourages the sharing of experience about the effects of work on health.

Body mapping helps raise awareness. Individuals may blame their symptoms on getting older, or being unfit, without realizing that others are being affected as well. They may accept the symptoms as 'part of the job' without realizing that they could develop safer ways of working if they were just to put their heads together and think about it.

Individuals may not realize that their aches and pains are linked to the job. Their GP may provide medical treatment but offer no occupational health insights.

For safety reps, body mapping is a valuable tool because:

- it helps workers to establish the link between their work and their health;
- it raises the collective awareness of safety issues among the people they represent;
- it helps them to identify clusters of common problems and their causes;
- it encourages workers to think about possible solutions to problems.

Online tools such as the 'Hazards Detective' are available on the Hazards Web site at: www.hazards.org.uk/tools.

How to body map

The body maps show front and back views of the body (see Figure 6.1). Using coloured pens or stickers, workers engaged in broadly the same kind of job are encouraged to mark on the charts where they suffer pain or injury while they are working.

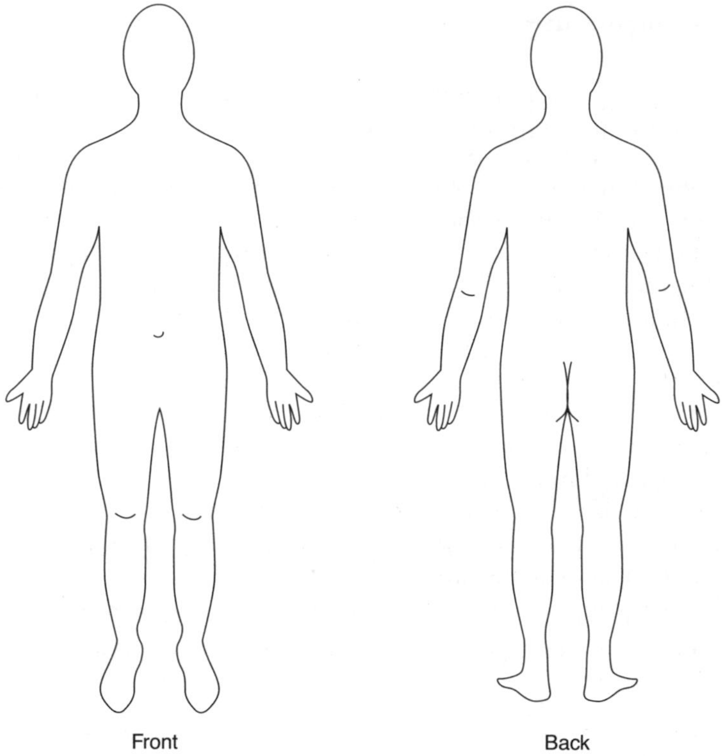

Figure 6.1 Body maps
(Source: *TUC*)

One technique is to use different colours to identify different problems, for example:

- red for aches and pains;
- blue for cuts and bruises;
- green for illnesses (stomach upset, dermatitis, etc);
- black for any other problems.

USDAW's advice is:

- Get members talking to one another about their own experiences.

- Make sure that as many people as possible in similar jobs take part.
- Ask members to write down the causes of the pain or injury beside the mark.
- Encourage discussion on ways in which the injuries could be prevented.
- At the end of the exercise clear clusters may emerge showing that many people doing the same job are suffering similar symptoms. This evidence, along with suggested causes and solutions from the members, can be taken as the starting point for negotiations with management.

Questionnaire surveys

Targeted questionnaire surveys can help to identify clusters of musculo-skeletal disorders among groups of employees. The technique itself can be applied to many working environments.

In the USDAW case above, the safety reps decided to carry out a survey of the 83 employees in the store, basing their questions on the types of complaints received – see questionnaire.

Questionnaire of checkout staff

CONFIDENTIAL

1. Male ☐ or female ☐ Please tick as necessary.
2. Please tick your age group: 16–20 years ☐ 21–30 ☐ 31–40 ☐ 41–65 ☐
3. Please tick how many hours you work: 0–8 hours ☐ 9–16 hours ☐ 17–24 hours ☐ 25–32 hours ☐ Over 32 hours ☐
4. Please state your height: .... feet .... inches.
5. Do you experience any discomfort on the new tills? Yes ☐ No ☐

6. Do you experience any pain working on the checkouts? Yes ☐ No ☐
7. Please tick any of the following boxes where you experience pain: neck ☐ shoulder ☐ forearms ☐ wrists ☐.
8. Have you consulted your GP about it? Yes ☐ No ☐.
9. Are there any other comments you feel should be made about the new checkouts?

   ....

(Source: *USDAW*, 1998)

Results

- Most of the employees surveyed (80 per cent) were female, with a majority working 20 hours a week.
- Over half the group were less than 5 feet 4 inches tall.
- A clear link was made between the height of the workers and the level of pain suffered. Shorter workers were worst affected by the new checkout designs because of the extra stretching and lifting involved.
- Over 30 staff (37 per cent) had consulted their GP.
- Over 90 per cent had pains in the shoulders.
- 80 per cent suffered pains in the neck.

Workers said they were stretching more on the new tills, particularly due to the extra distance to the conveyor belt, and the height of the keypad. Other complaints included twisting to reach the cash desk or carrier bags, and dissatisfaction with the chairs.

The union's report concluded: 'The high incidence of shoulder and neck pain, particularly in shorter workers, and the consistent comments about stretching, all suggest there is a problem with the physical ergonomics of the check-out stations.' The company should 'look again at the physical dimensions and adjustability of the new tills' to make sure that operators, especially shorter staff, were not having to work outside safe stretching limits (technically known as their 'reach envelope'). The union learnt of similar complaints in four other stores where newly designed checkouts were installed. The results were presented to management and, after some further consultations, the design and adjustability of the tills were radically improved.

Hazard mapping

Hazard mapping helps to identify and prioritize hazards. A rough sketch map of the workplace is used for a floor-by-floor assessment of where the problems are found. A TUC guide suggests that hazard mapping, like body mapping, gives a much more 'visual' picture than data from surveys alone. Corridors, stairways and outside areas should be included. In practical terms, the three techniques, of body mapping, surveys and hazard mapping, complement one another.

You need a simple floor plan of the office, factory or worksite, and a checklist of the main types of hazards you are likely to encounter (see page 29): physical hazards (noise, temperature); chemical hazards (cleaning agents, fumes); biological hazards (bacteria, needlesticks); work design hazards (workstations, working alone); and stress hazards (workload, harassment, shift-working, working hours).

The TUC sees this technique as an opportunity for workers to participate in creating a hazard map. The results form the basis of discussion among employees and safety reps, identifying the key issues to raise at health and safety committee meetings with management.

Working safely in a supermarket

Over 250,000 people work on supermarket tills, full or part time. One study found that the average checkout operator lifts the equivalent weight of a small car over the scanner in a single shift. By law, employers should provide a safe system of work, and training in safe working practices at checkouts.

But, according to Doug Russell, health and safety officer at USDAW, 'Many stores are so busy that even the big companies will have a good written policy at national level, and all the training materials, but in the stores themselves, is it carried out? Are staff told how to adjust the seat to their needs, and how to scan items safely?'

USDAW's 'safety partnership' agreement with Tesco is one of many management/union initiatives supported by Government funding, aimed at 'revitalizing' health and safety procedures and practices at all levels in the organization. Covering 100,000 USDAW members, it is the largest private sector safety partnership in the UK.

At Tesco, some of the first benefits of a new 'safety group' structure, covering all the company's 600-plus outlets, include: 1) joint staff/management safety forums in all stores – reps get training in health and safety, team building and problem solving; 2) joint consultation on new checkout designs, with the assistance of ergonomists, to reduce musculo-skeletal disorders. Local safety reps and the union's health and safety officer are represented on the project team.

Checkout risk assessments

The Health and Safety Executive's 'good practice' advice for checkout staff, safety representatives and shop managers states bluntly: 'Checkouts vary a great deal. Some are well designed and suitable, others are older and may be inefficient or unsuitable.'

The HSE provides the following checkout advice for managers and employees:

- **Provide suitable seating.** This may be a chair or specialized sit–stand stool. It must have adjustments for height and backrest, be easy to move and have a footrest. Make sure there is enough space to allow a comfortable working posture. Make sure the seat is in good repair, and that adjustment mechanisms work.

- **Face the workflow.** Cashiers should *face across the flow of goods.* Belt delivery and belt-plus-sliding-chute at the checkout are preferable designs. Scanners are preferred over cash tills. The 'reach envelope' to all equipment should not exceed 350 millimetres.

- **Variety.** Avoid long, unbroken stints. Regular, short breaks help offset the effects of fatigue. Rotate tasks regularly, eg with shelf filling, tidying up, office work, kiosk, restaurant and other work opportunities.
- **Weights.** Avoid items weighing over 5 kilograms if the cashier is seated.
- **Equipment.** Replace faulty bar code equipment.
- **Train** in good practice, eg scan with both hands, take regular breaks before fatigue sets in.
- **Encourage early reporting** of symptoms. This is likely to limit their severity.

Unions offer the following additional advice for checkout workers

Information:

- **Be informed.** Equip yourself with the information you need about how to do your job safely. This will prepare you for the risks you may be facing. Do this before symptoms occur. Ask your supervisor / line manager for a copy of any good practice guide issued by your employer.
- **Union guidance.** If you are a union member, get a copy of your union's manual handling guide.

Before you start work:

- Make yourself comfortable at the checkout before you start.
- Adjust your seat so you do not have long or awkward reaching to do.
- Ask for a footrest if you need one, so that your feet are firmly placed on the footrest or the floor.
- If you have problems with any equipment or if some items are too heavy, tell your supervisor, or approach your workplace safety representative.
- Make sure you receive training, so you can do your work safely.

Working safely:

- Do:
  - Use both hands to handle goods.
  - Keep the flow of items to be scanned running parallel to your shoulders.
  - Bring items to be scanned as near to your reach as possible.
  - Reduce awkward hand and wrist movements while scanning and weighing goods.
  - Take a break. Stretch your shoulders and back. Tension in the shoulders, neck and back restricts the natural blood flow to muscles and tendons. This increases the likelihood of a tear or strain injury, either from a sudden or awkward movement, or from using tired muscles too long without a break.
- Do not:
  - Twist your body as you lean across to pack a supermarket trolley.
  - Stretch and lift goods, eg to place them on a weighing machine.
  - Replace a basket or weigh goods by twisting whilst sitting.
  - Reach round to pick up carrier bags.

Further information

Body Mapping: Telling where it hurts, Health and Safety in Shops and *Charting Back Pain: A survey of women's experience*, USDAW, available from tel: (0161) 224 2804.

Hazards magazine (tel: 0114 276 5695; Web site: www.hazards.org.uk/tools).

Health and Safety in the Retail Sector, GMB Publications, Health and safety unit (tel: 020 8947 3131). A4 leaflets, free; larger booklets, price on request.

Available from the Health and Safety Executive:

Checkouts and Musculo-Skeletal Disorders, a free leaflet for employees available from HSE Information Line (tel: 08701 545500), free. Quote reference number: INDG 269.

Musculo-Skeletal Disorders in Supermarket Cashiers (1998), HSE Infoline (tel: 08701 545500).

Sources

Partners in Prevention: Revitalising health and safety in the workplace (2000), TUC, available from tel: 020 7636 4030, £30.

7 Safe and healthy call centres

Health and safety issues lie at the heart of the 'sweatshop' image of call centres. Employers say that some bad apples have affected the whole basket. But a reputation built on widely publicized 'Victorian' rules, such as putting your hand up to go to the toilet, is reinforced by their mass employment, with up to 2,000 workers under a single roof.

The call centre 'industry' is now a major employer. Up to 400,000 people work in 7,000 call centres in the UK, more than the combined total for car manufacture, steel and coal mining. And, as call centres diversify into 'multimedia' contact with their customers, offering access by e-mail, fax, a Web site, old-fashioned letters as well as the phone, they are more accurately called 'customer contact centres'.

The call and contact centre industry is today one of the few unionized growth areas in the 'new economy'. This is because many call centres emerged in already well-unionized sectors, like the energy and gas utilities, local government and banking. Tackling the wide-ranging health and safety issues in this sector is in the interest of workers, unions and employers.

Modern office, with a difference

Any modern office will present you with the same range of health and safety issues as working in a call centre: the use of desk, screen, keyboard, mouse, voice and headset. What probably distinguishes call centres from the rest of the world of office work is the *combination* of desk/screen/voice/headset/mouse used under *pressure*. The interplay of these hazards is not well understood. Added to this problem, the lack of control staff have over the pace and flow of work contributes a great deal to work-related stress and dissatisfaction.

For evidence of stress, look no further than the staff turnover and absenteeism figures in some call centres, with up to 80 per cent of staff leaving in a single year in the worst cases.

In response, the HSE issued Guidance for employers and safety inspectors in December 2001. *Advice Regarding Call Centre Working Practices* offers practical advice on display screen equipment; musculo-skeletal disorders; eyesight; noise levels; voice damage; work-related stress; daily and hourly monitoring; and environmental issues like air quality. Here, we look at:

- RSI in call centres: risk factors;
- work safely: adjustable equipment, breaks, stretching, eye care;
- advice for RSI sufferers;
- noise and hearing: acoustic shock, headset hygiene, ear waxing, hearing loss;
- dealing with abusive calls;
- stress.

The next two chapters cover: 1) voice care and voice loss (see page 99); and 2) self-directed work teams – the benefits of genuine teamwork (see page 105).

RSI in call centres

Aches and strains in the hands, wrists, arms, shoulders and back are the main cause of long-term sickness absence in the call centres. Most of the compensation claims faced by call centre employers in an HSE study were from workers suffering from musculo-skeletal disorders. The most common conditions were repetitive strain injuries. Pain or swelling in the hands, wrists or shoulders, and stiffness or pins and needles brought on by work could all be symptoms of RSI.

Employers have a statutory duty to carry out a risk assessment of any hazardous task or equipment, and consult their workforce through union-recognized safety reps (see Chapter 2, page 33). This applies to call centres as it does to any other workplace. Some employers are good at it; some focus on specific problems, such as the safe use of DSE equipment (computer/keyboard). Others do very little. This is one reason why 'individual solutions' – things you can do to work safely – are limited where your employer is not committed to a safety culture.

Static posture

'Sitting in the same position for seven and a half hours a day, looking at a video screen, I actually freeze into position. I have a bad back because of it. There is never enough time to get up and walk around, or take a stretch away from the desk' (Derek, call centre worker, Leeds).

Risk factors

The HSE identifies five main risk factors associated with RSI:

- repeating the same sequence of movements many times an hour/day;
- having to work very fast;
- having to use appreciable force;
- lack of control over the order of tasks or working methods; and
- working in awkward or tiring positions.

But high stress levels at work can also contribute to RSI. Not being able to control the pace of your work, the order you do it in and the way you work all contribute to stress and tension. In turn, working under these pressures increases problems associated with 'static posture', ie sitting in one place for long periods. The muscles tighten up when you are under stress. Continuing to work intensively under these conditions will increase the likelihood of developing RSI-type symptoms. Some types of 'referred pain' originate in tightened and sore muscles in the neck, but are felt in the hands, fingers, wrists and forearms.

New ways of working, such as teamwork and giving individuals greater control over their work, lead to higher levels of job satisfaction and can reduce stress at work (see Chapter 9). And, by reducing work-related tension and pressures, they also reinforce other steps you take to work safely, eg with a well-adjusted workstation.

Work safely

The essential requirements for the safe use of your workstation include the following:

- **Adjustable equipment**. You must be able to adjust your desk, screen, chair, document holder, footrest (optional). Figure 7.1 shows how to set up a workstation. Getting the 'ergonomics' right is the first step towards working safely with workstation equipment.
- **Still uncomfortable?** Much office equipment is still built for a standard-sized man, and is not downwardly adjustable for women or smaller people generally! If you find that this is the case, you need to explain the problem to your supervisor or safety rep.
- **'Hot-desking'** is a common working practice in call centres. If you share your desk, readjust it before you log on. You could enter a DSE checklist on your screen and follow it through before you start work (see box, page 89).
- **Taking a screen break**. The HSE says that employers 'should allow their employees time to take breaks away from the workstation and telephone, and the accompanying mental and physical stresses'.

The Display Screen Equipment Regulations 1992 (see box, page 88) require employers to ensure that any regular DSE users take adequate breaks. The HSE's advice says that both the length of break times and their frequency are equally important. 'Breaks' can range from doing other non-computer tasks to completely stopping work. The Regulations do not define 'break time' in terms of minutes every hour.

As an example of good practice, the RAC call centre, Bristol, allows regular DSE users to have a 10-minute break away from the screen every two hours.

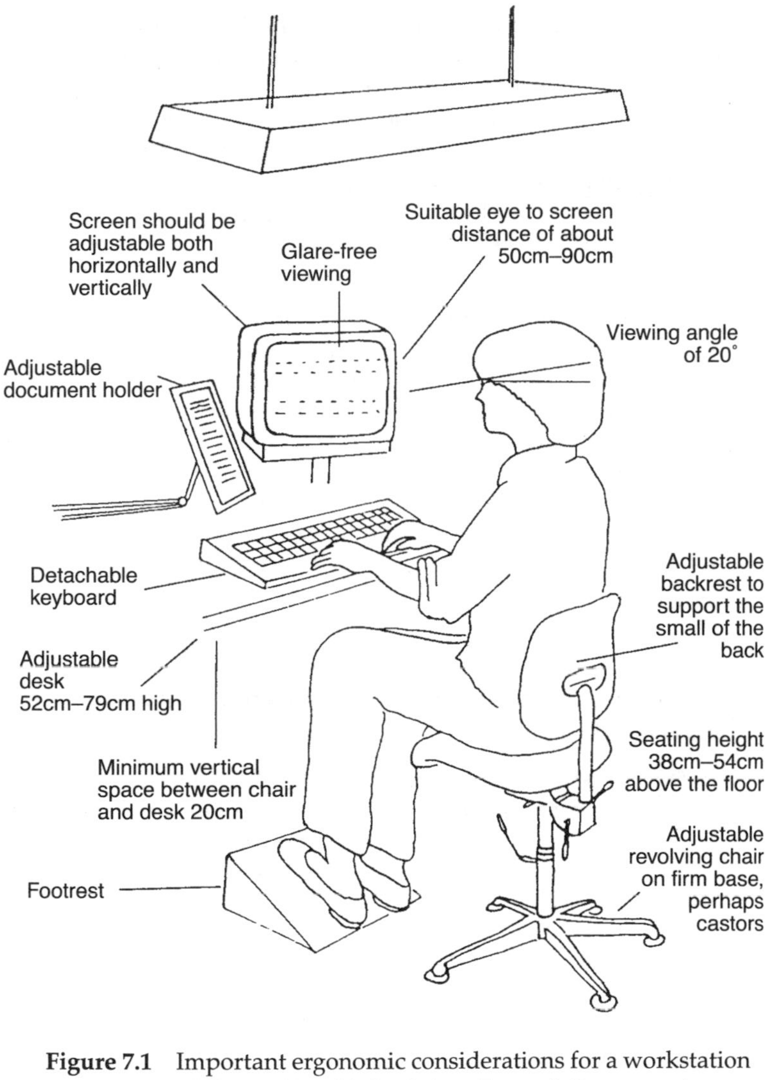

Figure 7.1 Important ergonomic considerations for a workstation
(Source: *Office Health and Safety,* Labour Research Department)

Take exercise breaks after a prolonged spell of any small, repetitive actions, such as typing and use of a mouse. Stretching aids blood flow, and helps break up the effects of static postures. Guides to 'office exercises' include side-to-side stretches for the neck, arm and body rotation, and downward stretches to the feet. See, for example, *The 9–5 Check*, available from The West End Physiotherapy Clinic (tel: 020 7734 6263) and *Back in the Office*, from BackCare, 16 Elm Tree Road, Teddington, Middlesex TW11 8ST (Web site: www.backcare.org.uk), price £1.50.

In its call centres study, the HSE did not find a single employer that encouraged regular eye tests. Nevertheless, employers have a duty, under the Display Screen Equipment Regulations, to offer regular VDU users eyesight tests when they start work with the organization and follow-up tests at regular intervals, and to pay for the cost of corrective spectacles.

'Blink rates' drop when using a VDU. Your eyes are less well lubricated, and are likely to become dry and sore. It is important to remember to maintain a normal blink rate when using a VDU. Frequent rest breaks, and work away from a VDU, give the eyes a chance to recover.

The Health and Safety (Display Screen Equipment) Regulations 1992

The Display Screen Equipment Regulations set out employers' legal duties, which include:

- Analyse workstations and reduce risks. Employers must look at:
  - the whole workstation: desk/computer/keyboard/mouse/document holder/other furniture/work environment;
  - the job being done; and
  - any special needs of individual workers.

Where risks are identified, employers must take action to eliminate or reduce them.

- Ensure workstations meet minimum requirements, such as adjustable equipment (desk, chair, keyboard), suitable lighting.

- Provide regular breaks or changes of activity. The Regulations require 'breaks or changes of activity', but do not specify the length of the break, or frequency. The general principle is that breaks should be short and frequent, rather than longer and infrequent.
- Arrange, and pay for, eye tests, and provide spectacles if special ones are needed. Further tests should be arranged and paid for at regular intervals.
- Provide health and safety information and training.

See also *Working with VDUs*, HSE: www.hse.gov.uk.

RAC Motoring Services: advice for VDU users

Do you regularly use a VDU for long periods at work? Do you want to avoid uncomfortable side effects? Then follow these simple guidelines:

- **Step 1**: **Adjust your chair** height to a position that allows you to place your feet flat on the floor. Ask for a footrest if your feet are off the floor when you have selected the best position to use your work equipment.
- **Step 2**: **Adjust the angle and height of your backrest** so that it adequately supports your lower back. Then adjust the tilt of your seat so that your body is inclined slightly forwards. This has the effect of making your body want to sit upright, and keep your back straight.
- **Step 3**: **Adjust your screen position**, so that your eyes look down on it from an angle of around 15 to 20 degrees (slightly below eye level).
- **Step 4**: **Adjust the viewing angle of the screen**, so that reflections and glare are minimized. If this proves impossible, you may need a screen filter.
- **Step 5**: **Adjust your keyboard**, so that your desk supports your wrists and forearms, and the angle feels comfortable during usage.
- **Step 6**: **Arrange your desk and equipment** to minimize the amount of bending, twisting and stretching involved in your work tasks.

- **Step 7: Take periodic breaks.** Get up and walk around, flex your fingers and stretch your legs to stimulate blood circulation and relax tense muscles. If possible, get right away from the workplace during your lunch break.
- **Step 8: If you spend a lot of time reading** from source documents or copy typing, place a document holder beside the VDU screen, at the same comfortable viewing distance from your eyes.

(Source: RAC Motoring Services)

Advice for RSI sufferers

You may reach a point where the work-related pain or discomfort you are suffering from cannot be ignored. You return to work after a few days off, and the symptoms are still there:

- Visit your GP. Tell the doctor about the type of work you do, and why you believe it may have contributed to your condition. Check Chapter 21 for what to expect from your doctor.
- Report the symptoms to your line manager, and record the incident in the accident book.
- Inform your workplace safety rep, if available. Discuss the possibility of further medical and legal advice from your union (see also Chapter 24 on legal action).
- In the short term, you may need time off work. When you return, you may need at least a temporary transfer to lighter duties, backed up by an agreement, preferably in writing, covering hours, pay, job description and review period for your revised duties (see Chapter 20).
- Learn more about your condition. Helpful advice is available from a wide variety of specialist sources, eg guides and booklets from the Labour Research Department and the RSI Association (see Further information, below) and BackCare.
- Contact your local occupational health project. (see Chapter 23).

Working environment

Dougie, an asthmatic, says:

> I have particular problems as an asthma sufferer because the terminals, keyboards and monitors are dirty, and not cleaned regularly. There is dust in the air all the time; the carpets aren't cleaned properly.
>
> The air is always dry and affects everybody's throats. You are constantly clearing your throat and taking drinks of water to keep it moist. Your nose dries up. The temperature goes up and down all through the night. One night you could be sitting there with a jacket on, the next you are taking your jumper off, it's so hot. I'm sure all this affects your health.

Noise and hearing

'Noise isn't just a hazard for factory workers using heavy machinery,' the GMB union points out in its handbook, *Health and Safety in Call Centres*. Both callers and call handlers raise their voice when using the phone. The key noise-related health issues in call centres are 'acoustic shock' and high background noise levels.

Acoustic shock

'Acoustic shocks' are freak soundbursts on telephone headsets, which can leave victims in severe pain. They can be deliberate, or accidental.

A BT employee, David Stonier, received £90,000 in compensation after suffering acoustic shock at a BT telephone exchange in Torquay. His union, the CWU, supported David's personal injury claim. He suffers from tinnitus, which is continuous sounds heard in one or both ears in the absence of external noise. These sounds, which may include whistling, ringing, buzzing or hissing noises, originate within the ear itself.

He says, 'I'm a lot more withdrawn than I used to be. I can be talking to someone in a supermarket, an alarm goes off, and I have to run out. It's always on your mind that something might set it off again and worsen it. Plus there is always the fear that it could eventually lead to deafness.'

The likely causes of freak soundbursts on headsets include:

- electronic sounds from fax machines;
- random electronic impulses;
- customers tapping into a handset; and
- malicious acts, eg customers blowing a whistle down the phone.

Background noise

The Royal National Institute for Deaf People (RNID) and the TUC commissioned a study of noise-induced hearing loss in a variety of workplaces, including call centres. Call centre workers reported that they are subjected both to acute intense noises and to prolonged high levels of sound through their headset. They felt that these noises caused either dulled hearing or tinnitus, or both.

The HSE suggests that if it is so noisy that you find it difficult to communicate with a fellow worker who is just 2 metres away, the noise is loud enough to damage your hearing. UK noise at work Regulations will be strengthened in 2005. Tougher new standards will bring noise protection to another 1 million UK employees, including call centre workers, bar and club staff, retail workers and teachers. For further information, go to: www.worksmart.org.uk/health.

Headset hygiene

Call handlers wear a headset all day, every day, so it is important that it is fully adjustable, checked regularly, repaired immediately, if necessary, and not shared. As the HSE's *Advice* points out, 'There is an increased risk of ear irritation and infection because headsets are worn so intensively.' Pooling headsets should be avoided, as it leads to the risk of cross-infection.

Ear waxing

Ear wax, produced by glands in the ear canal, cleans and moistens the canal. Usually, wax is produced in small quantities, and emerges naturally from the ear. Blockage can be caused by excessive secretion of wax, perhaps because the headset inhibits the clearing of the wax, or by an object, such as an earpiece, being pushed into the ear. Either cause is associated with headset use.

If you suffer ear wax accumulation, you should consult your GP, and consider informing your line manager and union representative. There may be a design fault in the headset.

In one call centre, a number of staff, all using headphones, reported a build-up of wax in the ear, causing hearing loss and discomfort. The occupational health nurse visited the headset manufacturer, who was aware of other cases. There was no sharing of headsets, eliminating cross-infection as the cause. Other explanations were explored: 1) a sensitivity reaction to one or more component parts of the hearing equipment; and 2) whether the headset had a single earpiece, inserted into the ear or hanging over it.

One employee was found to suffer from other allergic reactions, eg to cat hair, and was offered a different headset. For another worker, the solution was to use a single 'hanging earpiece', which sits over the ear, rather than in it.

Angry and abusive calls

> You're in the loop the whole time. If you have a bad call, someone is angry or where they have a problem you can't sort out, they will shout and swear at you. Yet there is no recovery time. This increases the stress on you. And if you press the 'make busy' button to disconnect yourself from incoming calls, so your phone isn't available for the next caller, someone will come down straight away and demand to know what's going on.
>
> (John, inbound call centre, Newcastle)

The Telephone Helplines Association (THA), in the training it provides for not-for-profit call centres and helplines, makes a clear distinction between handling angry callers and handling abusive callers.

The THA's *Guidelines for Good Practice* suggests that the policy of any service should be to allow call handlers the discretion to end an abusive call, defined as one involving sexist, racist or homophobic language, direct threats or swearing down the phone. Staff should be aware of the boundaries set by the service, and be comfortable with the standards. They should be confident that they will get back-up if a call is terminated.

Staff should receive sufficient training to feel able to handle an angry call. Techniques include not being defensive, acknowledging the anger felt by the caller, but trying to encourage the caller to move on to the service issue involved. The THA believes that you can generally work with an angry caller although, inevitably, some cases progress to abuse, and may need to be terminated, using the agreed guidelines.

Depending on the emotions involved in an angry or abusive call, staff may need time away from the phone to recover. In certain cases, eg with a particularly distressing call, counselling may be appropriate, or a debriefing conversation with a fellow worker or supervisor. Organizations should provide clear and understood guidelines for these situations.

Hearing, noise, headsets: what you can do

The following practical suggestions are based on the HSE's *Advice* and the recent TUC study of occupational health in call centres, *It's Your Call*:

- **Headsets**. Don't share your headset. It should be in good working order, cleaned regularly and adjustable.
- **Use noise limiters**. Headsets should be fitted with a noise limiter to prevent extremely loud, sudden noises from reaching the earpiece.
- **Microphone**. Use a noise-cancelling microphone: this ensures that only the worker's voice can be heard, thus *reducing background noise levels*. Position your microphone correctly: if your microphone is not set correctly, you have to raise your voice to

be heard by a caller, which puts strain on your voice, and raises the general noise level in the call centre.

- **If you are suffering** from a sudden noise shock, continuous sounds in your ear, short-term hearing dullness or wax accumulation, you should seek advice from your GP (see also Chapter 21), your line manager or occupational health service, and your union rep. Be sure to record the incident in the accident book.

Stress in call centres

> We sometimes make jokes about the feelings of faintness in our heads, and the pains in the chest. My colleague has exactly the same symptoms as me. We feel them at the same times, when you are really under pressure. You don't quite understand what's happening, or what the cause is; yet these are things you don't suffer from when you leave the workplace. The pressure of calls can be so great that it is difficult to regulate your breathing. There are no gaps between the calls. If I hang a call, I might be disciplined.
>
> (Dougie, call centre worker)

People experience stress in a variety of ways, ranging from emotional exhaustion, anxiety, depression and worry, to feeling burnt out, lacking enthusiasm and high dissatisfaction. Saying, 'I feel stressed' can cover all these conditions.

Dr David Holman, a researcher at the Institute of Work Psychology, University of Sheffield, studied job satisfaction and employee well-being among 500 customer service assistants in three banking call centres. He found the highest levels of dissatisfaction where work was monotonous, lacked variety, and where staff lacked control over the pace of their work.

On the other side of the coin, call centre staff experience much higher levels of job satisfaction when:

- their skills are well used;
- they have control over the method and timing of their work; and
- they can vary the type of work.

Dr Holman also found that call monitoring by supervisors listening into calls is the main cause of 'anxiety and emotional exhaustion' for call centre staff.

Poor job design

The Health and Safety Executive's Guidance says that 'poor job design' results from a combination of all these factors: limited tasks, low control over the pace and load of work, excessive supervision. These may all be inherently stressful. Yet most companies contacted by the HSE had a *reactive* approach to stress, 'treating' it once it had occurred, by offering individual support services, such as an employee assistance programme, counselling, lunchtime interest and fitness classes and evening social events.

Research shows that you can 'design out' a great deal of stress where workers enjoy higher variety of tasks, greater control over their work, more challenging work, and work together in a team in which they play an active part.

'Headaches, migraines, high staff turnover and high sickness absence levels were common. Most of the team leaders and call centre managers interviewed realized that call handling is stressful' (*Advice*, HSE).

Stress control: what can you do?

There is no one-size-fits-all solution to stress and job dissatisfaction in call centres. But tackling these three issues is a good place to start:

- **Genuine teamwork**. Providing workers with greater control through genuine teamwork, the solution adopted by a leading insurance company (see Chapter 9).
- **Call monitoring**. Asking if it is really necessary to monitor everything. Employers perhaps need to ask why they need so much information, especially when it is often seen as oppressive. Options include consulting with staff to agree alternatives, eg samples of work, at agreed times, with the results used beneficially to support employee development and learning.

- **Scripting**. Tightly controlled scripts are used to increase productivity, project company image and eliminate room for human error. But scripting is often seen as excessive and unnecessary. A simple alternative allows for flexibility around the core scripts that call centre staff are required to use.

A 'model' call centre

Industrial relations changed at BT's sales call centres after a nationwide strike in November 1999 over an oppressive management style and poor working conditions for agency staff. According to the CWU union, there was constant call monitoring, call handlers had to wrap up a call in 300 seconds flat, and if they did not, 'management would be on their back'.

A joint initiative to set up model working conditions includes:

- Call handling was converted into a group rather than an individual performance measure.
- Coaches supporting every team, offering support to call handlers with difficult queries.
- A 'People's Charter' involving zero tolerance of any form of harassment, bullying or intimidation; flexible working opportunities; and support for staff with work or personal difficulties.

Partnership initiatives

The Government's 'Revitalizing health and safety' strategy sees employer/union 'partnerships in health and safety' as a key way to reduce occupational ill health. Hugh Robertson, head of bargaining, health and safety at UNISON, describes these partnerships as 'about more than achieving mere compliance with safety Regulations, but breathing fresh life into the active development of safety practice'. Vertex Data Science in Bolton developed a partnership agreement with UNISON, including some key safety issues around rest breaks and a new safety committee. Elsewhere, Legal and General Assurance joined with MSF to develop new initiatives around its disabilities policy and safety training.

Just under half of the call and contact centres surveyed by Industrial Relations Services in 2003 recognize a trade union.

While partnerships and new rights to union recognition are creating a more positive climate for tackling health and safety at work issues, it is not all plain sailing:

> How did we grow the union? There was more stability on the nightshift; people tended to get to know each other better, and trust one another. That's how we grew the union. Really, it was all about health and safety issues: the stress, the call monitoring, and the humiliation of people. Pay came quite a long way down the line.
>
> (Dougie)

Further information

Advice regarding Call Centre Working Practices, Health and Safety Executive (tel: 08701 545500; Web site: www.hse.gov.uk/lau/lacs/94–1.htm), reference LAC 94/1, free.

Call Centres: The health and safety issues, MSF, available from tel: 020 7505 3000.

Diffuse RSI victory, July 1999, Unifi bulletin, available from tel: 020 8879 4261.

Guidelines for Good Practice, Telephone Helplines Association (tel: 020 7248 3388).

Health and Safety in Call Centres, GMB, available from GMB Health and Safety unit (tel: 020 8947 3131).

Holding the Line: A negotiator's guide to good employment practices in call centres (2001), UNISON, available from UNISON health and safety unit, 1 Mabledon Place, London WC1H 9AJ (tel: 020 7388 2366).

Office Health and Safety: A safety rep's guide (1999), price £3.25, and *RSI: A trade unionists' guide* (2000), price £3.15, Labour Research Department, available from tel: 020 7928 0621.

RSI Association (tel: 020 7266 2000; Web site: www.rsi-uk.org.uk). Publishes a regular newsletter, and can put you in contact with local RSI groups.

Working with VDUs, HSE (Web site: www.hse.gov.uk).

Sources

Call Centres 2003: Reward and Flexible Working, Industrial Relations Services, available from tel: 020 8662 2000.

Suff, Paul (2000) RAC rescues the call centre image, *Occupational Health Review,* November, available from tel: 020 7354 6761, price £12 (single copy).

8 Taking care of your voice

The larynx is the hollow muscular organ forming an air passage to the lungs, and contains the vocal cords.

'Dysphonia' – voice impairment – is an occupational health illness affecting workers in a wide range of jobs, from teachers and call centre workers for whom using their voice is an essential part of their job, to workers in noisy environments, such as sewing machinists, where you have to shout to be heard.

Julie McLean, a speech and language therapist, describes the voice as a 'bag of muscles' that can suffer from repetitive strain injury. She is a member of the Voice Care Network, a voluntary organization originally dedicated to providing teachers with information, advice and training on the care, development and use of the voice. The network now helps a wide range of workers. Anne, whose story follows, is one of them. This chapter ends with some practical voice care advice.

Answer that phone!

> When my voice started to break, I didn't put it down to the job at first. It felt as if I had a film of phlegm always at the back of my throat, that I needed to break by coughing. I was using my voice nearly all day, what with the calls I had to make, and talking to colleagues at work. It's something you do without thinking about it.

Anne, a customer service adviser in a telecommunications call centre, was receiving up to 40 calls a day from customers ringing to place an order for IT equipment, or making general enquiries about products. She would make up to 20 outbound calls a day. Any call could last up to half an hour.

'You obviously think it could be all sorts of things. But I had never smoked, and I worked in a non-smoking office. I went to my GP. He examined me, although he didn't ask about my occupation.'

Anne was referred to an ear, nose and throat consultant and, meanwhile, kept on working, even though her voice problem was steadily getting worse. The consultant examined her throat with a fibre-optic camera. But he found nothing medically wrong, and so referred her to an NHS speech and language therapist.

> On my first appointment she asked me straight away about my job, and whether I had neck or shoulder problems. She observed my breathing and how I used my voice. She took me through some breathing exercises, and, over the next few sessions, taught me how to use my diaphragm, and how to adopt the right posture when I was on the phone.
>
> I used to rehearse the various types of calls I had to handle, such as an angry call where you respond by tensing up, you lift your shoulders, which, as I found out, restricts your larynx. As I was going through that kind of call, she would watch, and helped me to relax my shoulders. Back at work I tried to remember the advice, to look right ahead, relax my body, calm myself down, which would in turn slow my voice down, and ease the use I made of it.

Because of the volume of calls she continued to handle at work, her voice was not recovering. By agreement, her therapist wrote to her employer, requesting a transfer away from call-handling duties. Soon afterwards, Anne met the company's OH nurse and, as a result, her work was reorganized for an initial one-month trial period, when she only received incoming calls meant for her personally, and processed customer orders in the back office.

People she worked with were very supportive, but the benefits only lasted a short while. When the organization reorganized its customer services, she found that she was again handling large numbers of inbound calls. Her team handled 100 calls in one 90-minute period. She had to continually remind her managers that she was not meant to handle large amounts of incoming calls. Eventually, she was permanently transferred to a back-office support role.

'In my annual appraisal, my manager enquired about my voice. I told her that it isn't as good as it was, and I mentioned the 100 calls we took in just over an hour. She just smiled and looked away. It was her decision to allow us to take on so much work.'

People don't think about their voice; they don't expect it to give them problems. But the voice is only a set of muscles. I'm much more aware of other people's voices now. I hear someone come into work with laryngitis, and ask them why they are at work. They say it doesn't hurt, but I know their voice sounds odd.

Voice care

Unifi, the banking and finance union, says that voice loss means more than just being unable to speak. Voice conditions are revealed by the following symptoms:

- pain, leading indirectly to a change in voice tone or quality;
- burning: a sore throat;
- tension: leading to change in voice quality;
- swelling of the vocal cords: leading to a harsh or deeper tone.

Voice specialists agree that consuming tea and coffee all day long can have a damaging effect on vocal cords. Caffeine is a diuretic, encouraging the discharge of liquids from the body. Like any other soft tissues, the vocal cords should be moist but are likely to become dehydrated. Anne says, 'When they become dry, you have to use more air pressure to move the vocal cords. If you are also not using your voice properly, so that it is already under strain, then banging dry vocal cords together for extended periods of time is likely to cause injury. It's like not having any engine oil in your car.'

What you can do for yourself

Julie McLean has run voice workshops for call centre workers. She offers this practical advice on preventing problems:

- Be well hydrated. Drink five or six glasses of water a day. Cut down on tea and coffee or, at least, avoid excessive caffeine.

- Stand or sit supported. Change posture as often as possible. Stand up and stretch regularly.
- Notice, and aim to release, excess muscle tension.
- Take time when you are speaking.
- Rest your voice when you have an infection.
- Seek help if your voice sounds or feels ill.
- Stop or cut down smoking.

She suggests you should beware of:

- Too much coffee.
- Forcing or pushing words out.
- Running out of breath towards the end of a sentence, and keeping going.
- Constantly clearing your throat. Drink water instead. The more you clear the throat, the more irritated it may become and the more you may feel you need to clear it again.
- Tightening abdominal muscles and upper chest breathing.
- Holding your head on one side for prolonged periods when using a handset.
- Thrusting your chin forward as you talk into a mouthpiece. Such a posture can put strain on muscles.

She advises: 'Pace yourself; try not to rush on to the next call.' Yet McLean acknowledges that call centre workers are expected to do just that to reach daily targets. The build-up of stress involved in hurrying on to the next call can tighten muscles and lead to inflammation of the larynx (laryngitis). 'And if you are rushing, then you tend to talk until you run out of breath.'

She advises voice conservation: walk up to someone to talk, rather than shouting across a factory floor. 'Sometimes, people do not realize how they are really feeling until the tension in their abdominal muscles, chest or larynx is released. I might offer to massage the laryngeal muscles if they appear to be tense. Or I might encourage other ways of speaking, such as encouraging someone to replace sharp expressions, like "What!" and "Now!" with gentler alternatives.'

One of the call centre managers whom Julie has worked with says, 'It seems obvious, but the human voice is one of the most used tools in a call centre, and it is one of the most fragile. A lot of attention is rightly directed at adjusting the display screen equipment and the seat. But arguably the most important asset to look after is the human voice.'

Steps you can take

If you are already a sufferer, then as a first step towards possible voice therapy, see your GP.

At work: report your condition

Whatever your job, you are likely to need at least a short break involving different duties. To begin with, you should:

- Inform your line manager, and record your condition in the accident book, especially if you need to take time off work.
- Make an appointment to see your work's occupational health department, if there is one.
- Contact your safety rep, if you have one. Your union rep can support you in meetings with management about possible adjustments to your work routines.

What your employer should do

Employers have a duty under the 'Management regs' (see page 35) to ensure safe systems of work, and to assess any risks faced by their staff. Prevention is crucial. Unifi argues that 'risk assessment should be seen as a key element in helping reduce the incidence of voice loss'. The union suggests the following negotiating points for safety reps:

- job redesign, including work variety, changes to the pace of work, provision of improved headsets if equipment is faulty;

- voice care training;
- regular work breaks away from the phones;
- environmental improvements, to air humidity, temperature and circulation;
- suitable alternative employment policy for employees suffering permanent voice problems.

Further information

Voice Care Network: a voluntary organization originally set up to provide support for teachers with voice problems, but now offering support to all sufferers. The organization produces a newsletter, *VoiceMatters*. Its coordinator, Roz Comins, says that voice damage is caused by the conditions you work in, and the pressures put on you. A *Voice Care Guide for Call Centre Managers* (also useful for frontline staff) and *More Care for Your Voice* are available from the Voice Care Network (tel: 01926 864000 or 01926 852933).

For advice on voice therapy, contact The Royal College of Speech and Language Therapists, 2 White Hart Yard, London SE1 1NX (tel: 020 7378 1200).

Visit the Unifi Web site (www.unifi.org.uk, go to research/health and safety) for practical advice and a 'model' policy on occupational voice loss.

Croner's Teaching Briefing, October 2000 (Web site: www.croner.cch.co.uk).

9 Teamworking in call centres

The call centre, located in the Thames Valley, showed all the obvious signs of a dissatisfied and stressed workforce: high turnover, high levels of sickness absence, reports of staff discussing their anxieties with their GP. Instead of improving customer service, the call centre was in danger of losing customers for the parent company, a leading insurance and financial services operation.

Janet, an MSF union rep, and Christine were working in different teams at that time. Christine remembers it well:

> After a few months working there you felt like a battery hen. Everyone was feeling stressed. You were just a number. You came in, logged on and you were straight on the phones. You had so many minutes to deal with an enquiry, so much time to wrap up the call; then it was straight on to the next one. You might still be wrapping up one call when another came through. Management seemed only interested in the statistics: how many calls had we handled. The stress was in the continuous work. You had no recovery time. And some calls could be very unpleasant.

Each team comprised 22 call handlers, but Janet says:

> It wasn't really a team at all. It was very lonely. You might make friends with people you sat near, but you didn't have the opportunity to build any relationships. We had monthly meetings, but they were top-down; the team leader would tell us what was going to happen.
>
> Control? We didn't have any. They knew when you went to the toilet, and how long it took. You felt frustrated, because you had ideas on how the call centre could work better, but there was nowhere to channel them.

Both women remember staff taking sick leave because of the pressure. Janet recalls:

> I was feeling tense, and snappy. It was all buzzing around in my head. I had to prioritize my feelings. This is a human being on the other end of the phone; I had to remember that. Unhappy callers would get abusive, but the rule was, you had to be nice to them. Sometimes, I would go home and not want to communicate with anyone. You could tell that some managers were also unhappy with the situation.

But there were no opportunities to feed these experiences back to management.

Getting critical

Staff turnover and customer losses were approaching critical levels. In response, management brought in a consultant occupational psychologist, with a wide brief to review work organization and survey staff attitudes. In one-to-one interviews, she explored the sources of stress and anxiety, including management attitudes and work control.

The results were presented frankly to team meetings. Janet recalls: 'Staff surveys showed they had to change their style, become more approachable and give us more say over our own work.' They proposed a radical change in work organization, based on the principle of 'self-directed work teams', and asked for volunteers to pilot the new ideas. Janet and Christine were among the uncertain few who took part.

Criteria for satisfying work

A three-day briefing for the 20 volunteers explored how teamwork based on self-direction would work in practice. Workshops focused on six 'criteria for satisfying work':

Related to the task

- Elbow room: the space that frontline staff have to make their own decisions.

- Learning: staff setting their own goals, and learning from staff and customer feedback.
- Task variety: the advantages of mixing different types of work, combining phone work with clerical tasks, training and developing new services.

Related to the team

- Support and challenge: working together, flexible working, covering for unavoidable absences, and a willingness to be a critical friend.
- Meaningfulness: staff knowing where their work fits in to the wider picture by handling the whole service provided to a customer, from beginning to end.
- Desirable future: the opportunity to develop, learn new skills and perhaps move on to another job in the wider company.

The volunteers were told that they would work in much smaller teams, with 11 members. They would decide for themselves how much responsibility they would take on for setting team and individual goals; deciding on work variety; managing absenteeism and performance; and resolving conflicts within the team. They were expected to meet weekly. Key tasks such as chair and absence monitor would rotate. Training would be provided in chairing skills, conflict resolution and key services.

Janet recalls their reaction: 'The older ones said, "Hallelujah!" The younger ones looked a bit scared. From the start, we realized that the two main aims were to improve customer service and staff attendance. We decided our own ways of tackling it.'

A review after six months showed distinct improvements in the number and quality of calls handled. 'We were quite shocked. But it came from the knowledge that we were in control of the work, and because we wanted it to happen. We were finally being allowed to do the job our way. So we did.'

Both women agree that the job is now more varied, more satisfying and less stressful, and staff attendance figures have improved. They identify these changes:

- **Mutual support**. 'We can now support each other. There is a lot of one-to-one coaching, instead of having to wait for a training slot. Working together releases time for training, or developing a new idea.' This is why customer service has improved.

- **Setting targets**. The team decides the average call time, currently eight minutes, with most calls to be handled within a time band around the average. 'It's what we decide, and feel comfortable with.'
- **Teamwork**. 'Self-management has improved everyone's involvement. We are now doing things like inviting experts from other departments to explain how their services work. We discuss ways to improve what we deliver.' They are learning to negotiate with management: 'We feel we have a certain amount of clout to say this is what we want; we need this to happen.'
- **Training**. There's a new training 'culture', with more people progressing to a higher grade. Teams now provide one-to-one training for new staff, whereas before they used to sit in a training area on their own, in what was called the 'cabbage patch'. Trainees are reaching the fully trained standard more quickly.
- **Absenteeism**. 'Attendance has been an issue for one or two people in the team. If necessary, it's discussed at the team meeting. The person leading on attendance will take it up. Being self-organized nips certain attendance problems in the bud.' But none of the self-managed teams has wanted to assume the formal role of taking disciplinary action.
- **Flexible working**. Shift swaps are organized on a one-to-one basis. 'Before, we had to go pleading to the centre manager.'
- **Rewards**. Financial reward for taking on more responsibility comes through higher individual performance-related pay increases, linked to improved skills.

Janet considers that the new working arrangements have not affected the union's essential role: 'We're still here to help when staff need it... The work is more satisfying. There's nobody that knows the job better than the people doing it.'

Further information

Holman, David and Fernie, Sue (2000) Can I help you? Call centres and job satisfaction, Spring, *CentrePiece* magazine (tel: 020 7955 7673), price £5.

Part 3

Stress, violence and bullying

Bullying at work

"I've got a problem with a colleague at work..."

People often begin slowly, when they want to talk about bullying at work. Bullying is very distressing and can be very destructive. It can happen anywhere.

What is meant by "bullying"?

"Bullying" can take many forms, for example:

* Physical violence, intimidation, name calling.
* Persisting in...picking on someone, criticising them in front of others, shouting to get things done.
* Insisting on making the decisions and on being right always – punishing anyone who offers ideas.
* Changing what has been agreed – or saying that it is not as agreed, even though it is – and blaming others for the mistakes.
* Giving someone work that you know they cannot manage, letting them fail to complete it, then blaming them.

People "on the receiving end" describe things that have happened to them, "The supervisor kicked a trolley at me. It hit my ankle. I was off work for a week."

"She has knocked me off my feet...stuck a pin into me as she has walked past."

"I am told off in front of everyone else and humiliated."

"The manager makes all the decisions and is always right. If you put forward ideas she makes you clean the stockroom."

"My work is always torn to shreds. If I pretend it's someone else's, there is never a problem."

What can I do about it?

First, realise that nothing can change unless you challenge it.

Second, gather evidence. Record what happens.

What next?

Find out, discreetly, what colleagues think. Maybe others are being bullied too. Realising that, may help you all and indicate the size of the problem.

Discuss the bullying with your trade union if you are a member.

Should I discuss the problem with the bully?

Yes, you may have to. Gather facts first about a number of occurrences. If more than one of you is being bullied see the bully together. You might also seek advice from your TU representative or ask them to be present.

Be assertive, but not aggressive. Be positive and calm. Stick to the facts. Describe what happened.

If insulting remarks have been the problem, say why they are unacceptable. If there has been disagreement about work instructions, tell the bully that you will write them down in future to avoid misunderstandings. If there has been violence say that you will complain to a senior manager, then do it.

Try to get agreement on better relationships in future. Record the meeting and continue to monitor events afterwards; then take it up at a higher level if necessary.

If the bully is an equivalent or subordinate colleague you might discuss the problem with your line manager.

If your own line manager's behaviour is the problem discuss it with him or her, otherwise with another manager.

Should I use the grievance procedure?

If your employer has a grievance procedure you and other colleagues should use that. If, later, you complain to an Employment Tribunal they will expect you to have done that.

Does my employer have to take action about bullying?

Your employer has a legal duty to look after your health and welfare.

Certain kinds of bullying can easily overlap with sexual or racial harrasment, which is covered by other legal provisions.

For more information visit the Acas website at **www.acas.org.uk** or call the Acas helpline on **08457 47 47 47**.

10 Tackling stress at work

The note she left behind read, 'I am finding the stress of my job too much. The pace of work and the long days are more than I can do.' In November, Ofsted inspectors had visited the school, Middlefield Primary in St Neots, Cambridgeshire, where Pamela Relf taught six- and seven-year-olds, and told her that her lessons 'lacked pace'. She had been teaching for 36 years. According to press reports, the inspection had a 'devastating' effect on Pamela Relf and her fellow teachers. The following January, she left her car near the River Ouse and, on a desperately cold day, walked into the freezing water. It made her heart stop.

A few months later that year, in May 2000, a teacher in Shropshire, forced to retire after suffering a complete nervous breakdown, was awarded £300,000 compensation in an out-of-court settlement.

Stress: the facts

An estimated 5 million workers suffer from high levels of stress at work, or one in five employees in the UK, according to a study funded by the HSE.

In a three-year enquiry, researchers at the University of Bristol interviewed 8,000 employees, using a 'conservative' cut-off point to define their level of stress. This report makes stress the second most significant cause of occupational ill health at work in Britain, after back pain and other musculo-skeletal disorders. High stress is associated with long hours, high demands, intensive work, lack of control over work flow, and unclear or inconsistent instructions, the study found.

With up to 13.4 million working days lost a year due to stress at work, the Health and Safety Executive (HSE) is piloting new guidelines for employers to tackle work-related stress, based on the *five steps to risk assessment* it applies to many other workplace hazards. Meanwhile, a ruling by the Court of Appeal in 2002 (*Sutherland* v *Hatton* and three other cases) has produced important guidelines on how the courts should deal with personal injury claims linked to stress at work.

In this chapter, we cover:

- defining work-related stress;
- 'fight or flight' reactions to stress;
- symptoms of stress at work;
- stress – the risk assessment approach;
- law on stress;
- women and stress;
- what you can do for yourself;
- taking legal action.

Defining work-related stress

This is how the HSE defines stress: 'The adverse reaction people have to excessive pressures or other types of demand placed on them' (*Tackling Work-related Stress*, HSE).

A certain amount of pressure is inevitable in any job, and can be enjoyable. Coping successfully with pressure can add to a sense of achievement, and help to motivate. But the danger of the 'good stress/bad stress' approach lies in the assigning blame to *individual workers* who fail to cope, rather than looking at the way in which *work itself* is organized.

'Stressors', such as excessive workload, long hours or impossible targets, are health hazards. This is recognized by the HSE, and is fundamental to the risk assessment approach to stress in its latest official guidance for employers.

'Fight or flight' reactions to stress

Stress is a natural *reaction* to excessive demand or pressure. When we feel pressured, hormonal and chemical defence mechanisms are triggered in the body. This is often called the 'fight or flight' reaction. It evolved so that we are better prepared to deal with dangerous or life-threatening situations. Mobilized for action, we begin to perspire, blood vessels to the skin constrict, muscle blood vessels swell, the stress hormones adrenalin and cortisol are released: adrenalin to accelerate heart rate, cortisol to promote blood sugar levels. We are ready to go!

Our bodies are well adapted to cope with short-term stress. It can even be enjoyable deliberately to trigger stress reactions in situations of our own choosing – in action sports, theme park rides, the performing arts.

But, if pressure is prolonged, too frequent or out of our control, the stress reactions in our body become chronic and can lead to ill health:

- Stress suppresses the immune system, increasing susceptibility to diseases. Tense muscles and soft tissues are more easily damaged under pressure.
- Psychological distress leads to depression and anxiety.
- Stress is linked with health-damaging habits, such as smoking, alcohol and escapist eating, all of which are associated with other diseases.

Symptoms of stress at work

Trade unions such as USDAW, the GMB and MSF have found it useful to classify stress symptoms into short- and long-term effects. Their stress-at-work guides show short-term symptoms as:

- anxiety;
- tension;
- irritability;
- forgetfulness;

- disturbed sleep;
- headaches;
- indigestion;
- weight loss / gain;
- skin rashes;
- muscle fatigue;
- raised blood pressure / rapid heartbeat;
- fall in performance;
- becoming accident-prone;
- increased alcohol, smoking, drugs;
- tensions at home.

Without an effective assessment of the causes of stress at work, and interventions to root out the key problems, short-term symptoms can develop into much more serious long-term ill health conditions. Union stress-at-work guides indicate the following long-term symptoms:

mental health	chronic anxiety depression mental breakdown suicide alcohol / substance abuse social isolation
digestion	diarrhoea vomiting peptic ulcers
immune system	lowered resistance to infections chronic asthma chronic dermatitis possible increased risk of cancer
heart and circulation	heart disease heart attack stroke hypertension

Stress – the risk assessment approach

In *Tackling Work-Related Stress*, the HSE recommends that employers tackle stress in essentially the same way as any other hazards at work, by using the *five steps to risk assessment* (see page 42).

Step 1: Identify the hazards

The HSE recognizes seven broad categories of 'risk factors' linked to work-related stress:

- The culture of the organization, and how it approaches work-related stress.
- Demands – such as workload and exposure to physical hazards.
- Control – how much say people have in the way they work.
- Relationships – covering issues such as bullying and harassment.
- Change – how organizational change is managed and communicated in the organization.
- Role – whether the individual understands their role in the organization, and whether their employer makes sure that they do not have conflicting roles to perform.
- Support, training and personal factors. These include training in the core tasks people carry out, support from peers and line managers and whether the organization caters for individual differences in approach, style and so on.

Almost any work factor linked to stress can be resolved by better work organization. Long or unsociable hours, a fast pace of work, workload, complex demands, lack of support – all can be dealt with, the HSE states.

Employers are encouraged to assess the extent of stress through staff surveys, staff turnover figures, absence data and other sources. Where unions are recognized, then identifying and assessing stressors should be a shared exercise, carried out through a joint safety committee.

Step 2: Decide who might be harmed, and how

The HSE suggests that different groups of staff might be affected by work organization, culture and demands in different ways. The key issue for employers is to discuss and consult with employees to find out where the immediate pressure points are.

Step 3: Evaluate the risk

For each of the hazards identified in step 1, employers are asked to answer three questions:

- What action are you already taking?
- Is it enough?
- What more could you do?

The HSE guide works through each of the seven risk factors in turn, explaining that employers have to adopt the same 'hierarchy of control' approach as for any other types of hazard. These control measures include:

- avoiding risks altogether;
- tackling risks at source, eg by reorganizing work to ensure people are clear about what is expected of them;
- adapting the work to the individual, and not the other way round, in such areas as work design and variety, allowing work control, and choice of working methods;
- develop a coherent overall policy on stress prevention, that covers the key areas of use of technology, work organization, working conditions, social relations at work (eg tackling harassment and bullying);
- provide clear procedures for people with grievances; and
- give clear and appropriate instructions to staff.

For example, *lack of control* is a key source of stress. It concerns the amount of say that individuals have in how their work is carried out. The HSE recommends employers take steps to devolve more control to staff by enabling them to plan and manage their own work, make decisions about how that work is completed and how problems should be tackled. The HSE says, 'Only monitor output if this is essential.' Alternatives include regular meetings with staff, and the creation of a supportive environment, where staff know managers will support them, even if things go wrong.

As we showed in Chapter 9, genuine teamworking can lead to much higher levels of satisfaction and employee well-being.

Step 4: Record the findings

Here, employers are reminded of their duties under the Management regs to record findings of risk assessments and share the results with staff.

Step 5: Review the assessment

At six-monthly intervals, or more often if changes are taking place at work, the risk assessment should be revisited, and the actions taken reviewed to see if they are still valid, or being enforced.

The TUC argues that stress should be treated as any other preventable workplace hazard, and urges safety reps to follow the HSE's risk assessment approach.

Piloting management standards

Building on its earlier advice to employers, the HSE has published a draft set of 'management standards' on work-related stress to be achieved through a new risk assessment process. Details of the approach it is piloting are at: www.hse.gov.uk/stress/stresspilot/standard.htm.

The management standards cover the six main factors that can lead to work-related stress:

- demands;
- control;
- support;
- relationships;
- roles;
- change.

For each standard, the HSE sets out a number of key stages, from analysis to action. The starting point is an employee survey to define the current state of play in each of the six areas. For the purposes of the HSE's pilot project, it assumes that one-fifth (20 per cent) of workers in any organization will say they are either very or extremely stressed by their work. This is based on the Bristol University survey referred to earlier. The HSE wants employers to take steps to cut this back by at least 15 per cent to a satisfaction rating of 85 per cent. Stress levels are measured by staff surveys and questionnaires.

For each of the six stressors above, the HSE defines what it calls 'the state to be achieved' by an organization if it is to minimize work stress.

Take the example of work demands. For 85 per cent of employees to say they are able to cope with the demands of their job, the HSE expects employers to:

- provide employees, including managers, with adequate and achievable demands at work;
- ensure that job demands match people's skills and abilities;
- ensure that employees in high demand have a say over the way they work, and receive adequate support from managers and colleagues;
- eliminate repetitive and boring jobs as far as is reasonably practicable;
- ensure that employees are not exposed to poor working conditions;
- eliminate exposure to violence or verbal abuse;
- provide opportunities for workers to raise concerns about health and safety, shiftwork and working time.

This practical approach is applied to each of the sources of stress.

Stress studies

Long hours, high work demands and other factors commonly associated with stress feature in many recent reports covering manual and white collar workers. Stress is not just a white collar issue. A TUC report, *Work Smarter: An end to burnout Britain,* reported that working hours in the UK are longer than anywhere else in Europe: an average of 44 hours a week, with 4 million people regularly doing more than five hours' unpaid overtime a week.

- A TGWU survey of bus drivers in Sheffield showed that shift patterns caused the most physical stress, affected sleep and digestion, and disturbed family and social life.
- A study of Whitehall civil servants (May 2000) found that having little influence and control at work is associated with poor mental health in men and a higher risk of alcohol dependence among women.
- Call centre researchers found that a high level of computerized work monitoring and its use to punish have negative impacts on employee well-being (see page 95).
- UNISON's survey of stress among manual workers (gardeners, home helps, refuse collectors) found the 'time available to complete tasks' to be an important source of stress.

The law on stress

There is no specific law against stress at work. The main *sources* of the duties requiring employers to tackle work-related stress include:

- **Health and Safety at Work Act 1974**. Under section 2 of the Act, employers have a general duty to ensure the health, safety and welfare of their employees, and this includes physical as well as mental health.
- **The Management of Health and Safety at Work Regulations 1999**. These Regulations require employers to undertake 'assessments of health and safety risks', and to take action to reduce or eliminate those risks.
- **HSE Guidance**. *Stress at Work: A guide for employers* says: 'Ill health from workplace stress must be treated the same as ill health from other physical hazards. Employers have a legal

duty to take reasonable care to ensure employees' health is not placed at risk through excessive and sustained levels of stress arising from the way work is organized, the way people deal with each other, or from the day-to-day demands.' Employers are expected to take full account of this Guidance on their policies and procedures.

- **Disability Discrimination Act (DDA) 1995**. To gain protection under the DDA, employees must have a 'disability', defined as a physical or mental impairment that needs to be substantial, long-term (over a year) and impact on their ability to undertake normal day-to-day activities. Employers are required to make 'reasonable adjustment' to working conditions of employees with a disability. Stress-related illnesses, such as depression or mental ill health will, in certain situations, secure protection under the DDA (for details, see Chapter 20).
- **Working Time Regulations 1998**. Britain's long hours culture is a major cause of stress. These Regulations limit the working week to an average of 48 hours, and set other minimum standards (see page 28). Some groups of workers, such as drivers, are excluded from the Regulations, while any individuals can 'opt out' of protection – see 'working hours' at: www.worksmart.org.uk.

Women and stress

The particular demands placed on women workers are often overlooked when people discuss stress. Most working women share the 'double burden' of juggling paid work with childcare and domestic responsibilities. They are also more likely to be working unsocial hours, and providing caring support for the elderly and disabled. Balancing these conflicting demands is difficult, tiring and stressful. And, as the GMB points out in *Working Well Together*, women are likely to work shifts and be subject to discrimination and harassment at work.

Women also tend to be concentrated in lower-paid, lower-grade jobs in the service sector, particularly retailing and the hotel and catering industry, where they have less control over their day-to-day work. Most nurses and teachers are women. The Bristol study shows these are two of the most stressful occupations in the UK.

Health and safety law emerged from the old heavy industries, in jobs traditionally done by men. Much of it is still all too necessary. But health and safety priorities also have to keep pace with the 'new safety and well-being agenda' in the new industries, such as call centres, and the rapidly expanding service sector as a whole. These jobs are done by women, by young people entering the workforce and, of course, by men displaced from traditional industries.

Accidents in the traditional industries could be sudden and deadly. Stress has been described as a 'slow accident'. But it still has the capability of causing serious or permanent ill health.

What you can do for yourself

If you think you recognize in yourself any of the individual stress symptoms we described on page 119, take that as a warning signal.

The following suggestions for individual action to tackle stress are taken from the National Stress Awareness Campaign, HSE Guidance, and advice from Mind, union and other sources:

- **Talk to someone**. When work gets on top of you, talk to someone you trust, at work or outside. It's not a sign of weakness, but taking responsibility for your own well-being. A few days later, perhaps talk to the person again, to see if you are any clearer about what you need to do.

- **Think things over**. Where do you experience the most stress in your life: at home or at work? Are you trying to meet impossible demands, by being the most talented, most capable person? Or, if you say these demands are imposed on you, where do they come from? The Amicus / AEEU's *Guide to Stress* and Unifi's guide, *Stressed Out?* both encourage you to listen to your own feelings, and beware of the warning signs. *Guide to Stress* includes a personal questionnaire listing a wide range of coping strategies, which you may, or may not, be using.

- **Look for warning signs**. The effects of stress on health vary considerably from person to person. But most of us have our own stress response, or 'fingerprint'. This might be headaches in one person, an outbreak of diarrhoea in another, losing sleep or loss of concentration in a third. The most important emotional changes to watch out for are *increases* in tension, irritability and moodiness. These changes may be more obvious to other people than to yourself.
- **Visit your GP**. Most GPs prefer to deal with stress by general advice or counselling (see Chapter 21). Just giving a full account of your situation can help you to focus on the likely causes of your stress, and possible ways forward. Ask about relaxation techniques. Time off work may help you to sort out some key issues.
- **Look after yourself**. Newspaper articles and union booklets often contain personal '10-point plans' for stress, including regular exercise, healthy eating, taking proper breaks at work and avoiding excessive caffeine. Try a creative activity (eg learning a musical instrument) or a new sport.
- **Reconnect**. Identify friends and family you really get on with and make definite arrangements to meet.
- **Avoid 'false friends'**, particularly increased drinking, smoking or drugs.
- **Try to regain some control**. Certain changes are in your own hands, while others will need to be talked through with fellow workers, a manager or your union rep. For example, if *workload* or work organization is the issue, discuss it with your supervisor. If you cannot resolve the problem there, talk with the personnel department, union rep or fellow workers. If *long hours* are the issue, build in a leaving time, say 5.30 pm, whenever possible. Try to work regular hours, and take holidays due. It helps to build your own support network of people you work with. To be able to sound off about the latest turn of events at work helps to bring it down to size.
- **Company policy**. Be aware of any company policies on harassment, bullying and racism, so that you know what stan-

dards the organization considers unacceptable, how to challenge the behaviour and what back-up there is.

- **Helplines**. Confidential helplines are operated by some major unions, employers (through employee assistance programmes) and national voluntary agencies dealing with issues such as bullying, mental health and a wide range of disabilities.
- **Returning after long absence**. Stress can lead to long-term absence from work. Getting back can be difficult, and requires your employer to make 'reasonable adjustments' to your former job (such as to your former workload, responsibilities or working time), or offer you a suitable alternative. You may be entitled to ask for adjustments to your former role under the Disability Discrimination Act (see page 220).

Taking legal action

There are three main options:

- To support a personal injury claim, you need evidence of injury or illness. You must be able to demonstrate that it was caused by your employer's neglect, with foreseeable consequences. And in stress cases you must *also* be able to show you suffered from a *clinically recognized psychological injury*, eg depression. The principles applied by the courts to occupational stress claims are the same as for any personal injury claim (see page 259). Remember, each case depends on its merits. There have been some well-publicized cases, with large pay-outs, but do not assume that all awards are easily won, or very large.
- If you have been working under pressure for too long it may be highly tempting to just walk away from the job and claim constructive dismissal (see page 173). But, as with personal injury claims, this is not an easy option guaranteed to succeed.
- If, after a long spell away from work, your employer fails to make 'reasonable adjustments' to help you return, you may also be able to take action under the Disability Discrimination Act (see page 220).

A Court of Appeal judgment in February 2002 (*Sutherland* v *Hatton* and three other cases) reduced the likelihood of employees successfully claiming damages, whilst reinforcing the need for managers to manage stress at work properly. The Court of Appeal laid down 16 'practical propositions' on stress that will influence future court rulings. These cover issues such as how far employers should be able to foresee stress among their staff, what kind of pressure is 'normal' and what employers need to do to avoid a breach of their duty of care. TUC health and safety specialist Owen Tudor was sceptical about one suggestion, that providing a 'confidential advice service' or helpline would let employers off the hook, since such services may be deemed to be inadequate. The need for employers to carry out a risk assessment is reinforced by the judgment, Tudor believes. As the Labour Research guide points out, the Court of Appeal decision is being challenged in the House of Lords, and stress cases continue to be taken by unions.

A stress at work policy

Stress at Work, a UNISON advice booklet for workplace health and safety representatives, includes a model stress prevention policy for negotiation with employers. It contains two key points that are found in many recent union agreements on stress at work: 1) **risk assessment**: making sure the employer gives priority to 'assessing the causes of stress, and introducing measures to reduce or prevent it'; 2) **support for individuals**: ensuring that the agreement provides for sympathetic support for employees becoming ill through stress.

UNISON's 'model' stress prevention policy

Parties to this agreement recognize that stress at work is a health and safety problem, and that employers have a duty under the Health and Safety at Work Act to take all reasonable and practical measures to prevent stress at work.

1. The Management of Health and Safety at Work Regulations 1999 require employers to assess health and safety risks, including stress, and to introduce prevention and control measures based on those risk assessments.

2. This agreement applies to all employees... regardless of age, gender, race, sexuality, grade or job.
3. Priority will be given to assessing the causes of stress, and introducing measures to reduce or prevent it. [Senior managers] will carry out these assessments in consultation with union representatives... training will be provided.
4. Where stress causes a deterioration in job performance, this will be treated as a health problem. The sufferer will be encouraged to seek help under the terms of this policy [with] no discrimination against individuals suffering from stress.
5. The HSE Guide, *Stress at Work*... will be used in determining the appropriate action to be taken by the employer.
6. Employees suffering from stress-related illnesses will be offered paid time off to attend stress counselling sessions... [provided] by independent and trained counsellors. This service will be strictly confidential between the counsellor and the member of staff...
7. Information and training will be given to all employees on the causes and effects of stress, how to seek help, and the arrangements for reporting causes of stress and work-related illnesses.
8. Where an employee becomes ill through stress, management will seek to identify the causes of stress, and eliminate them through changes to duties or working environment [or] employees will be offered alternative suitable posts, through agreed procedures for relocation...
9. This policy and its effectiveness will be reviewed regularly by staff unions and management...

(Adapted from UNISON)

Further information

Guide to Stress, available from Amicus/AEEU Health and Safety Department (tel: 020 8462 7755), free.

Surviving Working Life, available from Mind (tel: 0845 766 0163).

Tackling Stress at Work and *Working Time Regulations: A practical guide*, available from Labour Research Department (tel: 020 7928 0621).

Sources

A Practical Approach to Stress Assessment, Occupational Health Review (OHR) issue 103, and *Stress: The Court of Appeal guidelines*, OHR issue 101, available from tel: 020 8662 2000, price on request.

The Scale of Occupational Stress: The Bristol stress and health at work study, Smith, A P *et al* (2000), available from HSE Books (tel: 01787 881165), price £25.
Work-Related Stress in Manual Workers, Cox, Professor Tom *et al* (1996), available from the Health and Safety Unit, UNISON, 1 Mabledon Place, London WC1H 9AJ.

11 Violence in store

> I received no personal safety training when I was made assistant manager, just some advice on use of the CCTV and handling theft. The company was supposed to train the security guards we hired. But their pay was so poor that none of the ones we recruited had any experience. Anyway, none of them stayed long enough to be trained. Our guards weren't required to be trained before they started with us.

The menswear retailer, part of a national chain, is in inner Manchester. When Mark Palmer, the assistant branch manager, was attacked at night, the store had no security staff.

Mark's story

It was 12 June 2000. 'The store manager and I were working in the upstairs office when the floor supervisor put out a call over the tannoy, using a special code. In this case it was "Staff announcement, code nine". This told us that something was going on that we needed to deal with.'

Mark hurried down to the shop floor, and saw two young men, not shopping but behaving suspiciously. He decided not to approach them directly, and instead watched them from a distance.

> They came up to me and started to be very abusive, saying things like, 'Are you following us around? We're going to sort you out.' My manager joined me at this point, saw what was going on and backed me up. But they started having a go at both of us, saying they'd sort us out if we followed them around, or looked at them like that again.

Feeling quite threatened, Mark and the manager decided to back off.

> We didn't think it was worth our while getting involved. So we both went down to the front of the store to see if they left. I had a mobile phone on me, but didn't like to use it to call the police, in case they left the shop and nothing happened. Very often, if you back off, they will walk away, no problem.

Mark believes that his manager must have looked across to the men, because one of them came over and started to attack him. The manager fought back, and the men ran off, throwing boxes at both managers as they left. Mark then called the police, who came, made notes of the incident and advised both managers to ring again if there were any repercussions.

Because he was due to lock up the store at closing time, 9 pm, Mark told the manager that he was worried that the men might come back later. The manager accepted this might happen, and contacted head office to ask for some security support. But head office was not prepared to send any additional security cover.

Hammer blows

The manager went home at 6 pm, and Mark carried on working on the floor, helping refill shelves and managing staff. At 7.30 pm, he went out to the front porch of the store with one of the supervisors for a cigarette break.

> As I walked out the store, someone lunged at me and punched me hard on the nose. I was in shock. I didn't know where the person came from. I tried to walk back into the store, but he hit me repeatedly from behind, in the back and on the back of my head. I kind of made it to the rear of the shop. I don't know what was going on in my head. I just kept walking away. I didn't retaliate. I suppose I thought that once he had hit me, he'd go away. I managed to get myself to the office, and called the police.

One of them was using a hammer to smash the front glass. When he was being attacked from behind, Mark vaguely remembers turning round to see his attacker, and believes he might have been hit with the hammer. They smashed in the front of the store, and ran off.

Mark also rang head office, but again they were unable to provide security back-up. The police arrived after 15 minutes. Mark says:

> They wanted me to take a ride round the area with them, to try to identify the attackers. But I couldn't leave the store; I was responsible for seeing the staff off and locking up afterwards. If I had left the store unattended, I would have got into serious trouble. I was quite concerned about everyone else getting away safely, in case they were waiting for us later in the evening. The police couldn't guarantee being around – they said they were having a busy evening.

Mark had the mobile number of one of the company's store security detectives, who happened to be a personal friend. He came down to the store immediately. They rang an emergency glazier, and together spent the rest of the evening clearing up.

Mark's flat was above another shop in the same parade, where he lived with his wife. They were both anxious about returning to the flat, and stayed the night with her parents. He went to hospital the next day, and was diagnosed as suffering from mild concussion. His GP prescribed painkillers for the ache in the back of his head. Mark does not know for sure whether he had been hit with a fist or a hammer. But he does know that he was lucky not to have been more seriously injured. He was off work for six weeks altogether.

> During my time off I decided that I did not want to go back to that store, with the gang knowing who I was. I rang my union officer for advice. He suggested that I should write to head office personnel, letting them know my position, and ask for a transfer to another store. So I wrote a letter to the company, but I never had a reply. I phoned the store manager once, and sent in my sick notes. He never got back to me.

No 'victim support' here

Some days later, Mark and his wife returned to the flat. The police rang him soon after their return to say that they believed they could identify the individuals who had attacked him, and asked if he wanted to prosecute.

> I asked the police if the company was intending to take action, either on my behalf, or for the damages. No, the company wasn't going to prosecute, they said. I explained to the police that I had just returned to my home, and that if I did decide to take legal action, the gang would soon know where I live. The police understood, and gave me a couple of days to think it over.

Mark decided against taking action. Nobody from the company contacted him about their decision. 'And, of course, when I applied to the Criminal Injuries Compensation Board, for compensation for the assault, I was turned down because I was expected to formally report the crime.'

Supported by union solicitors, Mark is considering making a personal injury claim against the company.

> I want someone in the company to take responsibility. A couple of days after the incident, I went past the store, and they had four security guards on duty. But they should have provided adequate security at the time.

Preventing shop violence: the law

The Health and Safety at Work Act 1974 (HASAW Act) requires employers to protect the 'health, safety and welfare' at work of all their employees (see Chapter 2). Official Guidance from the HSE deals specifically with the issue of violence faced by shopworkers and managers. The local council is the enforcing authority for health and safety in the retail sector.

The HSE's Guidance, *Preventing Violence to Retail Staff*, sets out employers' obligations towards their staff. It covers victim support, clear incident reporting procedures, special leave to enable recovery and providing legal advice to staff. On no count did Mark's firm match the basic HSE standard.

'Freedom from fear'

A shopworker is attacked every hour of the working day, according to a study by USDAW, the shopworkers' union. The report, *Voices from the Frontline,* shows that workers have been pistol-whipped, stabbed with knives or broken bottles, and beaten with baseball bats. Nearly half of the workers surveyed reported physical assaults on staff in the previous 12 months.

In its 'Freedom from fear' campaign, USDAW aims to:

- Raise awareness of the issue of violence amongst union members and shopworkers generally, employers and the Government.
- Make workplaces safe. This involves using risk assessments to help identify any weaknesses in safety and security procedures, bringing in model safety policies and supporting workers at risk.
- Make retail crime a high priority for the police.

Advice in USDAW's safety guide for retail staff includes:

- Physical protection. Security measures, alarms, CCTV, protective barriers, panic buttons. 'Staff need to know how to operate the security systems provided.'
- Training. Staff need to know what they are expected to do when an incident develops, how to follow safe procedures and how to use security equipment.
- Be prepared. Would you know what to do if you spot a shoplifter or credit card fraud, or if there is a robbery?
- Cash handling. Cash should be kept out of sight, and not be allowed to build up in the till.
- Shoplifters. Over half of the physical assaults are linked to attempted shoplifting. Employers should make sure their staff know what to do if they see a suspected shoplifter. Employers should make it clear that no one should risk personal safety to protect property.
- Reporting incidents. Employers need to know exactly what risks their staff are exposed to. A reliable system of incident reporting helps to build the picture.

Further information

Preventing Violence to Retail Staff (ref HSG 133), available from HSE Books (tel: 01787 881165), price £6.95.

Violence to Staff: An USDAW guide, includes practical advice on preventive measures and a model safety policy. See also *Don't Be a Victim of Violent Crime* and *Late-Night Working,* available from USDAW (tel: 0161 224 2804; Web site: www.usdaw.org.uk).

12 Personal safety

'If this should happen...'

Diana Lamplugh describes the abduction of her daughter, Suzy, in July 1986:

> She had travelled round the world on the QE2, working as a beautician, and was extremely good as a salesperson. Suzy had coped with some very difficult clients, loved the work and thought she could do anything. So she comes back to London and decides to become a negotiator with an estate agent. She's very good looking; the job will give her independence. She's a real businessperson. She is 25 years old.

The estate agency was based in Fulham, south London, near her home.

> She thinks that it will be a great job and she's going to go for it 150 per cent. She walks in and says, 'I see you need somebody', and they say, 'As a matter of fact, we do. You look just the right sort of person.' This is how they did it in 1986.
>
> So she starts work and, because she's the only woman negotiator they have, because she is good looking, lives locally and knows people in the area, she sat in the window. Right in the window. And, sitting in the window, she hails people, invites them in for a cup of coffee, or arranges to meet them after work. So she brings in a huge amount of business.
>
> It never once occurred to her that she was being used. She never realized what they were doing, and never realized that it might attract someone who was very dangerous. She was, in a way, being put on a stage. People do that still in some workplaces, where both men and women are taught to be outgoing, to bring people in.

Diana's voice drops.

> She was ripe game. She was ripe game…
>
> But one of the things you always say to people is that if anybody goes out and about, leaves the office, they should leave a note saying where they are going, and where they are going on to next. And tell someone if they change their plans, because that's the most dangerous bit. More and more people now work away from an office.

Suzy left her office without saying where she was going.

> She left just a note – she was dyslexic, and didn't want people to laugh at her. Suzy had probably gone to view a property with a client. Since the new police investigation, we have no doubt that Suzy was abducted. She has been declared dead and assumed murdered. We are concerned that anyone in her risky position could have suffered a similar fate. We know much more now than we did at the time.

In 1986, Diana Lamplugh founded The Suzy Lamplugh Trust, the national charity for personal safety. Her energy and commitment have extended the influence of this campaign into many occupations and walks of life, and almost every aspect of personal safety and welfare. Diana says:

> One of the first lessons we learned in the Trust is that you cannot *tell* anyone how to be safe. The only way to enable someone to be safe is for them to *decide* to be safe for themselves. To ask questions of themselves, such as, 'What would I do if...?' or 'If this should happen, then what would I do?' Practising the 'If this… then what?' scenarios gradually becomes second nature. Knowing your way round gives confidence, changes behaviour and makes you less likely to be an easy target.

Violence may be unpredictable. But it is not unpreventable.

Here, we detail the scale of violence at work, the legal duties on employers, and what you, and trade unions, are able to do about it:

- a picture of violence;
- violence at work: assessing the risks;
- zero tolerance in the NHS;

- working alone: home care staff;
- personal safety: estate agents;
- what you can do for yourself;
- if you suffer traumatic stress;
- further information and advice.

A picture of violence

The number of violent incidents at work rose steeply during the 1990s. Nearly 1.3 million violent incidents were committed by members of the public against employees in 1999, according to the latest Home Office figures from its British Crime Survey (BCS). Nearly 3.5 million working days were lost during the year due to physical assaults and threats at work.

Yet less than a third of victims were either offered, or asked for, help from their employer. The greater the amount of contact with the public, the greater the risk faced by the employee. Some occupations face very high risks of violence. Preventing violence is a high priority for unions in the public and private services.

The 11 highest-risk occupations, according to the BCS, are:

- security and protection workers;
- nurses;
- care workers;
- public transport workers;
- catering and hotel workers;
- education and welfare staff;
- teachers;
- shopworkers;
- managers and personnel officers;
- leisure service workers;
- other health professionals.

The BCS says violence to healthcare staff is under-recorded because 'the victim may feel the offender was not entirely responsible for their actions', and so may not wish to report the incident as a 'crime'.

The risk of violence is much higher for women aged between 18 and 25 years. It is also highly prevalent among managers and supervisors. Responsibility for intervening to handle a violent situation often falls on their shoulders. Offenders see management as a 'softer' or more 'legitimate' target for abuse.

'Crunch' situations

The BCS says violent incidents were most likely when:

- someone providing a service decides to withdraw it from a customer, eg asking a customer to leave the premises;
- a customer or client is angry about some aspect of a service, eg the speed of delivery, a benefit cancelled;
- a customer is fraudulently trying to obtain goods or services, eg with a stolen credit card;
- shopworkers confront a suspected shoplifter; and
- handling customers who have drunk excessive amounts of alcohol.

Less than a third of victims were offered, or sought, help from their employer. 'Employers could potentially play a bigger role in offering victims advice and support', the Home Office points out. This is reinforced by the fact that half the assaults and a third of threats were by clients or customers known to the employee.

In total, 604,000 workers suffered at least one incident in 1999. Over 300,000 workers were assaulted on at least one occasion, and a further 340,000 had been threatened. But only one worker in five (18 per cent) had received any formal training or advice on personal safety in their current job.

Violence at work: assessing the risks

Employers have clear legal obligations to prevent violence and threats at work. These hazards should be tackled in the same way as any other risk faced by their workforce (see Chapter 3). The HSE defines violence to staff as: 'any incident in which an employee is abused, threatened or assaulted by a member of the public in the course of their employment'.

In its formal Guidance, *Violence at Work*, the HSE urges employers to adopt a 'straightforward, four-step management process' to control violence.

Stage 1: Find out if you have a problem

This includes consulting staff and keeping good records of incidents. The potential for violence to arise is assessed by:

- looking at the *jobs* people do;
- identifying the magnitude of the risk of violence; and
- identifying the frequency with which violence actually occurs.

This means that it is not enough to look at the tasks on their own. The vulnerability of the employee and the *circumstances* in which the work is undertaken should also be considered.

UNISON's guide, *Violence at Work*, highlights two key groups of risk factors: 1) job-related risks; and 2) risks in the work environment.

Job-related risks include:

- handling money;
- providing care to people who are ill, distressed, afraid, in a panic or on medication;
- relating to people who have a great deal of anger or feelings of failure;
- facing friends and family of clients, who may be concerned, anxious, angry or afraid, or feel inadequate faced with a large 'bureaucracy' from which they are seeking help;
- enforcing rules or regulations, eg benefit officers, council inspectors;
- providing an essential service, or having the power to withdraw it.

Risks in the work environment include:

- working in isolation: in clients' homes, in an isolated unit or at times when colleagues are off duty;
- working under pressure: in circumstances of staff shortage or heavy workloads, or in the absence of an alternative support that can be offered to a client;
- working in chaos: workplaces that are crowded or too busy, with public areas that lack seating, refreshments, a telephone or recreation area for children.

Surveys can be helpful. An USDAW union survey found that 12 separate shop robberies occurred between 5 pm and 7 pm. The results highlighted a key issue of the timing of attacks, and had useful security lessons for the level of security, staff protection and other procedures involved in late-night opening.

Stage 2: Decide what action to take

The HSE requires employers to decide who might be harmed, and how. Preventive measures – the changes needed to eliminate or reduce risks to acceptable levels – include:

Safe systems of work

- Safe procedures for home visits, interviewing clients or patients, handling abusive phone calls and handling cash.
- Working alone. Is it necessary to leave staff alone for long periods? If so, what training do they receive in the tasks they perform without supervision? Are they easily able to contact a supervisor, or the police?
- Reporting incidents: with clear guidelines requiring staff to report all incidents of violence, threats, abuse and 'near misses' to a defined line manager.
- Crisis management procedures.
- Post-incident support procedures.
- Education on the impact of violent attacks.

Safe workplaces

- Improving facilities for the public in waiting or queuing areas (telephones, refreshments, adequate seating, play facilities, reading material, information on queue times).
- Secure areas: limiting public access to public areas; ensuring that public car parks and access ways are well lit.
- Personal alarms/panic buttons.

Training, information and support

- Ensuring that all public-facing staff receive training in agreed personal safety procedures, and are taught how to manage specific situations that they face.
- Support for victims of violent or abusive incidents should include:
  - crisis support;
  - debriefing;
  - specialist counselling where required;
  - legal help (see page 259).

Stage 3: Take action

At this stage, employers are expected to implement an action programme arising from the issues identified in their risk assessment. This includes training on safe systems of working, raising the level of safety awareness and encouraging reporting of incidents.

Stage 4: Check what you have done

Regular checks should be made on working arrangements, through a health and safety committee and regular consultation with safety reps where a union is recognized. Employers should monitor incident records and, if violence is still an issue, take further action.

Zero tolerance in the NHS

The 'Zero tolerance zone' campaign against violence to NHS staff reflects a determined drive by unions, employers and the Government seriously to change attitudes towards violence in the NHS. In 1998, the Government set national targets to reduce reported incidents by 20 per cent by 2001, and by 30 per cent by 2003. All NHS trusts (the employers) are required to record all incidents directed against their staff, and to have clear strategies to achieve the target reductions in violent incidents.

Yet violence and aggression against NHS staff by patients and their carers or relatives costs the health service around £69 million a year, according to the National Audit Office (NAO):

- Violence accounts for 40 per cent of all work-related health and safety incidents reported in the NHS.
- The number of violent incidents increased to 95,500 in 2001/02, up 13 per cent on the previous year.
- Around 40 per cent of violent and aggressive incidents are not reported.

In May 2002 a *Nursing Times* survey on violence showed that four nursing staff in 10 had suffered a violent attack in the past three years. But two-thirds of nursing staff were not offered any follow-up support, counselling or training, and only 4 per cent were offered help in pressing charges.

According to *Nursing Times* deputy editor, Alison Whyte, the survey showed that 'Violence against nurses remains invisible because it is not being taken seriously by NHS trusts, the police or the government.' The magazine proposed an action plan including regular risk assessments wherever nurses work with potentially violent patients. Trusts should notify the police about each assault, and compile statistics on assaults. And nurses should be offered effective training in managing and de-escalating violence.

The practical application of risk assessment to potentially violent situations is well illustrated by the violence at work procedures adopted by the New Possibilities NHS Trust (see box, below). Employing over 1,200 staff, New Possibilities specializes in working with people with learning disabilities.

New Possibilities: violence at work procedure

- The Trust adopts the HSE's *definition* of violence and aggression at work.
- Risk assessment: must be carried out in working areas where violence and aggression pose a significant risk to employees. Recommendations are made to eliminate or reduce the risk to the lowest level reasonably practicable. The risk assessment should be documented, and reviewed annually, or sooner if circumstances change.
- Obtaining assistance. If violence erupts, or a threatening situation develops, employees should immediately contact their line manager/on call manager for assistance, and/or the police.
- Reporting. All incidents should be reported to the employee's line manager. The HSE must be notified if the employee is absent from work for more than three consecutive days.

Guidance for employees to prevent or avoid violence and aggression

- Where possible and appropriate, *create space* between yourself and the aggressor (known as a 'reactional gap').
- Remember to communicate, as far as possible, in a non-confrontational manner. When in direct face-to-face contact, remember to blink.
- Attempt to calm the situation by keeping your own temper – listen, empathize.
- As far as possible, keep yourself between the aggressor and your exit or escape route.

External visits

- Be on time. Arriving late can be inflammatory for someone who might already feel uncertain. Prepare for the visit, try to pre-empt questions and have answers prepared.
- Before leaving your base, ensure that you have complied with the guidance on external visits, ie leave all the relevant details, addresses, time out, estimated time of return, etc.

- Check for any known details about the person you are going to visit. If there is a known history, discuss the situation with your line manager before you visit.
- On arrival, park your car in a well-lit area. Keep car keys separate.
- The risk of harm may not necessarily come from the client, but from relatives or other people on the premises or in the vicinity.
- Remember that you must not put yourself at risk. If in doubt, get out, and report the incident to your line manager.

Interview room

- Check that a suitable room is available. Elect to use a room with a panic button, and ensure someone is on hand to respond if needed.
- Ensure someone else knows you are going to conduct an interview, and where.
- If the room has only one door, show the client in first, so that you end up nearest the door. You then have the advantage if you need to leave early.

Abusive telephone calls

- Be patient. After a few moments the abusive language may abate. If the caller does not calm down, advise him/her clearly that unless he/she is able to continue the discussion in a civil manner, the call will be terminated.
- If it is impossible to supply a satisfactory answer, offer to pass the caller to a manager, or take down the caller's number to return the call when the answer is known.
- If, after giving a warning, behaviour does not improve, then you may terminate the call.
- All calls of this nature should be mentioned to your line manager, and the incident noted.

Support services

- Employees who feel traumatized by an act of violence or aggression inflicted on them may obtain support from their line manager, the OH department or a counsellor.

(Source: Health Services Report)

Working alone: home care staff

> She was going to visit a client on a council estate. Someone, she has no idea who, attacked her, pushed her into a bin store and locked her in. She banged and banged on the door, and screamed, until she was let out. But when I told the care manager, she just laughed it off. She thought it was funny. That's one reason why home carers don't report it; they're made to feel it's part of the job.

Ann Browning's survey of unreported violent incidents revealed some real 'horror stories'. Soon after she was elected to represent the 250 home carers in her division, Ann attended a UNISON health and safety course, and came back with the idea of carrying out a telephone survey.

> I found that 95 per cent had gone back to the office and never said anything. Yet, they go to strange, unfamiliar and unsafe areas to visit clients. Girls are attacked in their cars, but they go home and don't say anything about it. Because of parking restrictions, especially in city centre areas, they often have to park and walk some distance to a client's home. They face violence from clients, or from people their clients live with. Many visits are part of a daily routine, so anyone can notice when they come and go. Because we wear a uniform, people think we carry drugs, although we don't.
>
> When I rang them they just talked and talked. The things they experienced made my flesh creep. They had never told anybody before. We do not work in an environment where we can talk to each other. Partly, it's down to the way our work is organized – we work away from the office most of the time, so we rarely get together to talk. But partly it is due to management's attitude. They don't want to know. Really, it is a kind of bullying. When I speak up, they think it must be me. But someone resigned because she couldn't stand it any more, and let it all out at her exit interview, about the risks she had faced, and how impossible it was to say so.
>
> They always say to me, 'Don't say I spoke to you'.

Few staff were aware of the council's personal safety policy.

> We are lone workers. The policy includes a violent incident report [VIR] form for any actual incident or occasion when you are in fear

of violence. But trying to get the carers to see these risks are *not* part of the job is another thing altogether. It's a real problem getting them to fill in a VIR form. My survey did it for them, in a way.

The survey led to management taking personal safety more seriously. Recent negotiated improvements include:

- two-way radios for home care staff;
- parking permits for use in restricted areas;
- lone working training days; and
- 'safer driving' training.

Ann says, 'It's a start'.

Personal safety: estate agents

The Suzy Lamplugh Trust received a call from a mother whose daughter had a Saturday job for an estate agent. The daughter was working alone in an office on a new housing scheme, taking large amounts of cash in advance, as deposits on properties. She had to carry the money home, on foot. It could be thousands of pounds. The Trust handles many such enquiries from anxious parents or employees. They ask, 'Is this right?'

Around 120,000 people work as estate agents and surveyors, in 11,500 local offices across the UK. Most offices employ around a dozen staff. The larger employers appear to take their responsibilities seriously. For smaller firms, this is often not the case.

The Royal and Sun Alliance group of estate agencies includes some well-known high street names such as Barnard Marcus, Fox & Sons, and William & Brown, most employing three or four staff per office.

A group-wide risk assessment led to safety improvements applied across all the UK outlets:

- mobile phones issued to all branches for staff attending appointments away from the office;
- new guidance on the use of 'distress codes'; and
- half-yearly health and safety checks of all local branches.

The National Association of Estate Agents (NAEA) has drawn up personal safety guidelines for estate agents with the help of The Suzy Lamplugh Trust, available from the national body.

What you can do for yourself

These suggestions for personal safety at work are drawn from the experience of voluntary organizations such as The Suzy Lamplugh Trust, trade unions and employers:

- Take time out to think about your personal safety.
- Make a list of hazardous situations where you have felt at risk, or were caused harm, during the course of your work. For each occasion, ask yourself these questions:
  - What happened and where? How did you react? Did you tell anyone? If not, why not?
  - Could it happen again?
  - What would you do next time?

Talk to people you work with

People who have suffered an abusive or violent incident at work often don't want to talk about it. They blame themselves, feel 'sorry' for the client or work in an organization where safety isn't taken seriously.

Nevertheless, you should discuss your main concerns with people you work with, your union safety rep or line manager. There may be many steps you can take for yourself.

Often, discussions at work turn to single ideas like panic buttons, personal alarms and self-defence training. These may all be essential, but together they may not be enough. Your organization is responsible for ensuring safe systems of work, and safe workplaces. Taking steps yourself needs to be followed by your organization also changing, to plan violence and abuse out of the workplace.

Golden rules

- Don't accept violence or threats as part of the job.
- Don't blame yourself.
- Assess your own risks.
- Take action to prevent them occurring.
- Review how your ideas are working out.
- Talk to other people at work, your union rep or fellow workers.
- Talk to your management about carrying out a workplace risk assessment.

Traumatic stress

Violent attacks, threats, verbal abuse and other uniquely traumatic experiences can leave you feeling isolated, shocked and confused. But the effects of prolonged bullying may be equally psychologically damaging. Occupational psychologist Noreen Tehrani argues that you are most at risk when bullying involves unrelenting criticism and intimidation, and unpleasant personal remarks, in situations where there seems to be no escape. Targets of bullying are rarely exposed to a single traumatic incident. But the mental ill health consequences of prolonged bullying can be as serious as those experienced by someone involved in a single event, such as a train crash, fire, rape or serious assault.

Tehrani suggests there are a number of common reactions to traumatic incidents. You may feel:

angry	tense	irritable
jumpy	emotional	restless
alone	guilty	tearful
anxious	down	tired
numb	worthless	depressed

You may relive the traumatic incident in flashbacks or nightmares. You may have problems with concentration, eating, sleeping or sex. All these reactions are normal and should begin to fade in a week or so after the incident. However, sometimes feelings do not go away, or they start at a later date. If so, you may need further support.

Post-traumatic stress disorder

Your feelings or behaviour may persist some time after the incident. The most common symptoms of post-traumatic stress disorder (PTSD) include:

- reliving aspects of the event: with nightmares, or feelings that the event is happening again, a kind of continuous 'action replay';
- avoidance: symptoms related to avoiding doing anything to remind you of the trauma, eg a person traumatized by an attack on the street may find it difficult to go out alone;
- heightened irritability: sleep disturbance, outbursts of anger, sudden shock reactions.

If this is the case for you, then you may find that 'debriefing' or counselling will help you to come to terms with the incident. Talking it through may help to consolidate the incident into your memory where it belongs, rather than constantly re-experiencing it as if in the continuous present.

A responsible employer will recognize the duty to support you. But support that employers offer ranges from offering virtually no help at all, to a comprehensive trauma care programme that begins at the time of the incident and continues for as long as is required.

If your employer's support is unhelpful, or non-existent, visit your GP. Explain what has happened to you, and how you feel. Before visiting your GP, you may find it useful to read Chapter 21.

It has been shown that 're-experiencing' the traumatic event in a controlled and safe environment can bring about a greater understanding of what has happened to you, and will help you develop your own coping skills. Debriefing gives you an opportunity to talk about the incident, to make your recollections more coherent and to gain a greater understanding of how you can deal with its impact. Your employer may offer this kind of support through an employee assistance programme or its OH service. There is no 'one-size-fits-all' formula to support individuals after a traumatic event. Employers' support needs to be flexible enough to recognize that individuals will cope in very different ways. Too much 'analysis' can, for some people, slow down their recovery. For others, debriefing may be really helpful.

Noreen Tehrani has developed a five-stage trauma care programme. The programme starts with immediate care and support for individuals after an event. This is followed by a voluntary 'debriefing', to work through the incident and understand what happened. The next steps include a further psychological debriefing to help reduce the emotional response to an event. Finally, the programme is evaluated to make sure the employee is getting better. During the process, the organization itself learns about traumatic hazards in the workplace, and the steps needed to prevent it happening again.

Noreen Tehrani suggests some helpful hints. You should:

- talk about your feelings;
- ask for help;
- speak to your GP;
- try relaxation to unwind;
- go back to work;
- take exercise.

You should not:

- over-consume alcohol;
- take unprescribed medication;
- cut yourself off from colleagues;
- get over-tired;
- skip meals;
- bottle things up.

Support from your employer

An employer adopting good practice would consider taking the violent person to court, and providing you with help and support when going to an identity parade or court appearance. However, employers' practices vary widely in these areas.

Legal advice

If you are injured at work or in connection with your work, or suffer ill health as a result of the job you do, you may be entitled to recover compensation from your employer. This could cover the injury itself ('pain and suffering'), and financial losses and expenses caused by the injury. (For information on obtaining legal help, see Chapter 24.)

Criminal injuries compensation

You may also be able to make a claim from the Criminal Injuries Compensation Authority (CICA). To qualify, you must have suffered a personal injury as a result of a crime of violence, or attempting to prevent an offence. 'Personal injury' includes both physical and mental injury or illness. The mental injury needs to be a medically recognized condition, such as anxiety, depression or post-traumatic stress.

The incident should have been reported to the police, and be serious enough to qualify for a minimum award of £1,000. Claims should be made within two years of the event. Compensation is assessed on three criteria:

- compensation for injuries;
- compensation for loss of earnings;
- compensation for special services, eg cost of care.

For further information and a claim form, contact the CICA (tel: 020 7842 6800 or the CCA helpline: 0800 358 3601). For further advice, contact your union's legal service, a CAB or law centre.

Victim Support, the national charity providing support to victims of crime, is contactable through a network of local schemes. They can be contacted via your local police station, or by telephone (tel: 020 7735 9166 (England and Wales); tel: 0131 668 4486 (Scotland); victim support line: 0845 3030900). Their Web site is at: www.victimsupport.org.uk.

Personal safety guidance for nurses

- Always keep calm and explain quietly what you want people to do and why.
- Avoid confrontation and always report all incidents and near misses.
- If you suspect that a patient is going to be difficult, try to take a colleague.
- If you visit a patient in his or her house, always check you have means of assistance – do not assume there will be a phone in every house you go to.
- Trust your instinct: if you are uncomfortable with a situation or surroundings, do not continue.
- Always check you know where there is an easy exit, so you can get away if you need to.
- Communicate your movements; always let other people know where you are at all times.
- Hospitals are big places so consider using alternative routes, especially at night.
- Think about situations and try to imagine how other people are feeling.

Further information

The Suzy Lamplugh Trust is the leading authority for personal safety. Resources include:

- Guidance sheets, including those on nursing, bullying, driving, stalking, car parks, minicabs, home alone. Free, but the organization relies on donations.
- *Living Safely* and *Violence and Aggression at Work: Guidance for employers*.
- Personal shriek alarm. An ear-piercing shriek will shock and disorientate an attacker, giving you vital seconds in which to get away. Price £9 (include £1.50 for postage and packing).
- Training and other resources.

Contact: The Suzy Lamplugh Trust, 14 East Sheen Avenue, London SW14 8AS (tel: 020 8392 1839) or visit www.suzylamplugh.org.

Bullying Trauma May Open New Legal Challenges, IRS Employment Review 753 (tel: 020 8662 2000).

Can Zero Tolerance Deliver?, Labour Research, May 2003 (tel: 020 7928 3649).

Personal Security Guidelines, National Association of Estate Agents (tel: 01926 496800).

Preventing Violence to Retail Staff (ref HS (G) 133), price £6.95, and *Prevention of Violence to Staff in Banks and Building Societies* (ref HS (G) 100), price £6.50, available from HSE Books (tel: 01787 881165).

Protect nurses from violence, *Nursing Times*, May 2002.

Violence at Work: A guide to risk prevention, UNISON (tel: 020 7388 2366).

Violence at Work: Findings of the British Crime Survey, Budd, Tracey, www.home-office.gov.uk / rds / pdfs / occviolencework.pdf.

Working Alone: A health and safety guide, UNISON, available from UNISON health and safety unit, 1 Mabledon Place, London WC1H 9AJ.

Working Alone; *Health and Safety and Home Care Staff*; *The GMB Guide for NHS Workers*; and *Health and Safety for School Support Staff*, available from GMB health and safety unit (tel: 020 8947 3131), free.

13 When bullying goes unchecked

Maxine's story

If her case had been handled differently, within agreed anti-harassment procedures in her workplace, Maxine's complaints about bullying need not have led to her devastating loss of self-confidence, nor to real financial hardship.

When, in autumn 1998, Maxine James started her new job as medical secretary at a hospital in central Scotland, she had almost 30 years' experience of office work. She worked for a consultant rheumatologist and his team of doctors. For a salary of just under £11,000 a year, her role included handling correspondence, typing reports, liaising with local GPs, arranging patients' appointments and their transport, and a great deal of filing.

She liked the new job at first, and got on fairly well with the woman she shared an office with, Pat.

> She had been there 13 years, so she felt she was my senior. Actually, for a long time I really did think I worked for her.
>
> I like to take people as I find them, so I didn't really bother too much about comments that other people made about her, that she could be a bit of a troublemaker. But then gradually things started to go wrong. She became verbally abusive, saying things like, 'There's nobody in this hospital likes you.' 'You're a moody bitch.' 'Nobody's got a nice word to say about you.' But when it came to shouting at me, she was very clever; she wouldn't say a word out of line if anyone else was around.
>
> She realized she could bully me. And I just let her, perhaps because I am from the old school. You get on with your work; you don't complain. If you've got problems, you should get out. So I didn't mention it to anyone else at work.

There were arguments over sharing their small office. Pat refused to share the space equally. There was so little room for Maxine to store her files that her husband would help her after work to sort and stack them for the next day's work.

> The verbal abuse gradually became worse and worse, until I was bringing my troubles home. I was fighting with Mark, my husband, rowing with my girls. It was so awful that one day I asked my consultant [Dr X] if I could move offices. He refused, because the only available office was a quarter of an hour's walk away, in another building.

It appeared to Maxine that Dr X took Pat's side.

> Perhaps it was because she had been there longer. He didn't seem to want to rock the boat with her. After more than a year of this, it became so difficult that I decided to speak to my union rep. I didn't want to give too much away. But he knew exactly who I was talking about. Without really telling me, he met Dr X and my line manager, to explain the problem.
>
> I thought, well, I have to stand up for my rights eventually, when it's starting to affect my home life. Finally, Dr X allowed me to move to another office, but he was not happy about it. It was as if he had been backed into a corner, and had no choice but to move me because I had involved the union. But, I was able to get on with my work; it was great.

Dr X's attitude towards Maxine changed abruptly. He began to describe her as difficult. He changed her work routines for the worse, expecting her to make additional daily visits to his office to collect and deliver work, even though each round trip was taking up to half an hour.

Meanwhile Maxine was also working for other doctors in a team under increasing pressure of work. They would often make urgent demands for case notes, X-rays or blood test results. And there was always a large amount of filing.

She began to fall behind with her work. Locating a single set of case notes and delivering them to the clinic might take the best part of an hour. She asked the typing pool supervisor for some clerical support, and her boss if she could be excused the lunchtime visit. Each request was refused. Instead, her boss increased the number of required visits to four times daily. She would waste up to five hours a week hurrying between her office and the clinic. She often missed lunch and tea breaks, worked late and started going into work early.

An attempt on her part to explain the difficulties she was facing led to a row. She says, 'I asked him when I was supposed to do my own work, having just spent two hours on the phone talking to patients, trying to sort out problems with transport, appointments, you name it. Telephone work was a big part of my job, liaising with GPs, following up patients' cases who hadn't received appointments.' He argued that she 'knew what she was getting into' when she moved office.

The last straw was an argument with the typing pool supervisor over a booked holiday. Maxine says, 'If I hadn't already been feeling as bad as I was, I would probably have thought nothing much of it.' Her boss had approved a single day's leave and signed her holiday card, which she forwarded to the supervisor to record. But when she returned to work, the supervisor accused her of taking unauthorized leave. When Maxine objected, she was called 'nothing but a liar'.

Maxine stuck it out for the rest of the week, and by the end of it she had decided to quit. But a friend at work in whom she had confided persuaded her to talk to one of the other union reps, Leslie. She was persuaded not to resign, but instead to go and explain everything to her GP.

She told her GP everything, about the situation at work, about putting on weight and feeling 'generally awful'. The doctor wrote 'stress reaction' on her sick note, and signed her off work for two weeks. Maxine was to be away from work for six months, from March to September 2000.

On the breadline

The stresses at work, often aired at home, and now financial hardship, put her marriage under great strain. After two months' absence, her service-related sick pay expired. Her husband, who is registered disabled, was also not working. Maxine says she 'lived on chocolate and crisps, which were a lot cheaper than buying a meal'. Her youngest daughter took on three cleaning jobs to help pay the bills.

After three month's absence, she had her first appointment with the hospital's occupational health (OH) service. The OH doctor was anxious to get her back to work, and arranged for her to see a stress counsellor. After two sessions, the counsellor said she would inform the physician that, in her opinion, the stress was work-related. This confirmed her own doctor's assessment.

After five months' absence, she and Leslie, the shop steward, had their first meeting with personnel. Management initially claimed that there were no other suitable jobs available. But, when the union rep found this not to be true, Maxine was redeployed to a secretarial post in another department, well away from the sources of stress at work. She says:

> I've lost all my self-confidence. I've been a secretary for 30 years. But if someone looks at me the wrong way now, I will turn away and cry. I blame myself for what happened. I should have been stronger. Before, when my daughter was being bullied at school, I went there and sorted that out. I know what it's like; I was bullied at school, too. I find I am getting very forgetful. I developed a stutter at one point. I was very distracted by it all.

Recently, in a chance conversation with a colleague, she learnt that her former boss had rowed with Pat. Dr X had told her friend, 'Now I believe it wasn't Maxine's fault.'

Union tackles bullying

Leslie, the shop steward, often discussed Maxine's case with her branch secretary in the hospital.

> I was leading on Maxine's case, but we would often discuss it. When I told her about management's difficulties with finding another job for Maxine, she called personnel and asked to be faxed all the clerical vacancies. They were just trying to get rid of Maxine by being as obstructive as possible. We went back to personnel with details of the vacancies, and Maxine was offered one of them! We got all her current terms and conditions transferred to her new position.

'The Trust has to take responsibility. She has suffered so much financially and emotionally, through no fault of her own. What I admire Maxine for most is that she is willing to do something about the bullying she suffered.'

UNISON conducted a bullying survey in the Trust. Over 300 responses to a confidential questionnaire revealed widespread bullying, harassment and intimidation, mostly by senior managers and consultants, or by colleagues on the same level as their targets. The union is using the evidence to negotiate stronger anti-bullying procedures, and to ensure that staff and managers are better informed of its contents.

The union is currently dealing with 10 similar cases of bullying, but so far few people have been willing to come forward and make a formal complaint, out of fear of losing their job.

> I advise them to go to their GP, explain what's going on. The GPs write that it's stress-related. I encourage them to keep a diary. They may be off for a fortnight, come back to work, feel a bit stronger; things are OK for a while; then it starts all over again. More people need to confront bullying here.

Maxine's case: some lessons

Maxine's case is far more typical of the messy and unhappy experience of bullying than big-style compensation cases highlighted by the media. Lessons from the way Maxine's case was handled include:

- **Being aware of the right procedure**. Maxine had not read the Trust's anti-harassment policy, and was not aware of both the informal and formal stages, and the availability of support from a 'designated officer' whose role is to deal with complaints and offer support and counselling.
- **Management ignored the warning signs**. Dr X, her manager, also failed to operate the procedure set up to deal with exactly this kind of situation. In line with procedures, he ought to have interviewed both parties, dealing with the problem informally, and promptly, at source. He appears to have ignored the *warning signs*, such as signs of tension between the two.

- **Dealing with the 'target', not the bully**. Maxine spoke to her union representative, again informally. She did not want to be seen as a 'troublemaker'. So, understandably, the union rep spoke 'informally' to her manager. This time, Dr X reluctantly agreed to allow the target, Maxine, to move offices. He appears to have held a grudge against Maxine from that time. Meanwhile, the *bullying issue wasn't addressed* at all. With hindsight, one could suggest that the union rep was at fault in not facing management with its responsibilities under the anti-harassment policy.
- **Union support at work** Another shop steward, Leslie, persuaded her to visit her GP. Her doctor diagnosed work-related stress, and she began an extended period of sick leave. This *support at work* helped to save her job.
- **Long-term absence.** After three months' sickness absence, the Trust's OH service invited Maxine in for an interview, under its 'absence management' procedures. It is not clear why the Trust waited so long before contacting her. *Typically, the longer the period of sickness absence, the more difficult it becomes for someone to return to work.*
- **Stress caused by bullying**. Counselling sessions confirmed that her absence was linked to stress, and the main cause of that was bullying.
- **Union support for return to work**. Maxine was entitled to be offered suitable alternative employment under the organization's return-to-work procedures for long-term absentees. This was not forthcoming. Union persistence helped overcome some early difficulties, both over vacancies and unsuitable offers. She was eventually made a reasonable alternative job offer.

Supported by her union, Maxine is now undertaking a formal grievance against her employer for failing in its duty of care towards her, and breaching its own anti-harassment procedures.

14 Tackling bullying at work

This is an official definition of bullying: 'Prolonged conflict between individuals, including bullying or where staff are treated with contempt or indifference' (*Employers' Guide to Stress at Work*, HSE).

Destructive Conflict, the UK's largest-ever study of workplace bullying, found that one in ten people were bullied at work over the past six months, or around 2 million of this country's 24 million employees. As many as one in seven current 'targets' of bullying say they are being victimized on a daily or weekly basis. For many sufferers, bullying is a long-drawn-out affair, lasting well over a year. It involves high levels of mental ill health, depression and anxiety.

This evidence shows that bullying is one of the most significant *hazards* affecting people at work today. The joint authors of the study, Helge Hoel and Cary L Cooper, report that nearly half of the 5,300 people surveyed had been bullied, or witnessed bullying taking place, in the previous five years.

Yet employers have clear legal duties to tackle and deal with workplace bullying, harassment and intimidation, whether by supervisors, managers, fellow workers or the public. And you don't have to put up with it, as our self-help advice shows.

In this chapter, we look at:

- the effects of bullying on individuals;
- where bullying flourishes;
- what you can do for yourself;
- keeping a diary;
- bullying at work: the law;
- Cath Noonan's story: 'I know a bit about the way they work'.

The effects of bullying

Bullying not only makes you ill. People who are bullied often lose their self-confidence, sometimes long term. Their self-esteem can be shattered; they are at risk of suffering from work-related stress. And stress, in turn, has strong links with ill-health effects, both physical and mental. Effects on the individual include the following (*Bullying at Work*, Amicus–MSF):

Physiological

- headaches/migraine;
- sweating/shaking;
- feeling or being sick;
- irritable bowel;
- inability to sleep and loss of appetite;

Psychological

- anxiety;
- panic attacks;
- depression;
- tearfulness;

Behavioural

- becoming irritable;
- becoming withdrawn;
- increased consumption of tobacco, alcohol, etc;
- obsessive dwelling on the bully, and seeking justice or revenge.

As Helge Hoel points out in *Victims of Workplace Bullying*, 'Psychologically, bullying frequently manifests itself in depression, anxiety and nervousness'. Bullying is likely to lead to problems in concentrating at work, irritability and insecurity.

Where bullying flourishes

Many organizations fail to distinguish between 'managing' and 'bullying'. The personality 'defects' of a bully, such as aggressiveness, sarcasm, anger and maliciousness, flourish in certain work cultures:

- a highly competitive environment;
- organizations undergoing radical change, or serious cuts;
- in a climate of insecurity, eg of redundancy;
- under 'tough' and hierarchical styles of management;
- where there is little staff participation or consultation;
- where there are excessive demands on people;
- where there are no procedures to tackle harassment or bullying.

An anti-bullying policy at work

The HSE recognizes that bullying and harassment are common sources of stress. In its advice to employers, *Stress at Work,* the HSE expects organizations to operate effective procedures to deal with conflicts between employees. This includes bullying and racial or sexual harassment. Employers must investigate complaints thoroughly and operate a grievance procedure. In other words, employers must take steps to deal with poor *relationships* at work, which can be a major source of stress.

The HSE began piloting a set of stress 'management standards' in 2003, covering the six factors that can lead to work-related stress:

- demands;
- control;
- support;
- relationships;
- roles;
- change.

Further details are available at: www.hse.gov.uk/stress/stresspilot/standard.htm.

As regards work relationships, the HSE has set out certain minimum conditions for employers to be able to achieve acceptable standards. These include:

- Clear procedures to prevent, or quickly resolve, conflicts at work. These procedures must be agreed with employees and their representatives, and help employees to report in confidence any concerns they have.
- A policy for dealing with unacceptable behaviour at work. This must be agreed with employees and their representatives, and widely communicated in the organization.
- An approach to the way teams are organized that ensures they are cohesive and have a sound structure, clear leadership and objectives. Individuals in teams should be encouraged to be open and honest with each other, and be aware of the penalties associated with unacceptable behaviour.
- Procedures that encourage staff to talk to their line manager, employee representative or an outside counsellor about any behaviour that is causing them concern at work.

Union policies

Unions generally prefer to negotiate a specific policy on bullying. As the UNISON guide points out: 'The normal grievance procedure will not always be sufficient, as the facts of the case need to be established in a sensitive way, and the bully may be a line manager, who is normally the person a problem is raised with in the first instance in a grievance procedure.'

Key sections of any anti-bullying policy include:

- have commitment from the very top;
- be jointly drawn up and agreed by management and trade unions;

- define what is acceptable behaviour, and what is not;
- recognize that bullying is a serious offence;
- recognize that bullying is an organizational issue;
- apply to everyone;
- guarantee confidentiality;
- establish clear informal and formal procedures for dealing with it;
- guarantee that anyone complaining of bullying will not be victimized;
- commit to using a risk assessment approach to bullying.

The most difficult part for employers is owning up to the possibility that bullying is widespread, rooted in the organization's culture, and is about 'macho management' styles.

However, they should review their organization in a systematic way so that they understand in what circumstances bullying can arise and determine the steps needed to reduce that risk, such as management training, better lines of communication, an anti-bullying culture and an open procedure allowing staff to talk about their concerns.

Unions such as UNISON and MSF use workplace bullying surveys to help identify bullying 'hot spots'. Other sources of information include employee attitude surveys, and explicit questions in exit interviews.

What you can do to tackle a bully

Remember this basic point: whatever steps you decide to take, you have the right to work in a safe environment, and this includes not being harassed or bullied by anyone. You should not have to suffer in silence.

Confronting a bully yourself is not easy. It's probably only effective in its early stages. These suggestions are drawn from the experience of people who have suffered from bullying, and from union and voluntary sector advice:

- **Talk to someone you feel you can trust**: a friend at work, your union rep, a personnel manager or someone outside the organization. Talking to someone helps overcome the feeling of isolation experienced by many targets of bullying. Some employers nominate 'harassment advisers' to support and provide confidential advice to victims of bullying. They receive special training in their role, understand the procedures, and keep contact with the victim through their complaint. This helps overcome the uncomfortable fact that the line manager is often the bully, or that he/she lacks the skills or sensitivity to handle the problem adequately.
- **Contact The Andrea Adams Trust helpline or The Suzy Lamplugh Trust.** Tell the person you speak to what is going on. Say how it is affecting you, even if it seems trivial. Get a copy of the TUC's guide, *Bullied at Work? Don't suffer in silence.* Some unions provide a dedicated antibullying helpline. Unifi offers self-assertion training courses (see Further information, below, for details).
- **Try talking to the bully**. If you decide to approach the bully yourself, then first of all go over with someone you trust what you want to say. Have a clear idea of what you would expect to happen. Then, tell the bully how you find his/her behaviour unacceptable. Concentrate on the behaviour and its effect on you. Make a note of this meeting, who said what and the outcome. If this *informal* approach doesn't work, there are other *formal* options to use later.
- **Keep a record or diary**. Record the date, time and place of important incidents, details of abuse, accusations, changes to your job. Write down your feelings at the time and your own response. Collect evidence. Keep a diary, and a copy of letters, memos and appraisals.
- **Make a formal complaint**. If you decide to make a formal complaint to your line manager or personnel department, be well prepared. Familiarize yourself with your employer's procedures or guidelines on harassment, bullying and sexual or racial harassment, if any. Get a copy of your job description if you believe your harassment includes changes to any of your key responsibilities.

- **Employee helpline**. Some employers and trade unions operate confidential advice lines, or counselling through an employee assistance programme. You could consider accessing one of these services.
- **If you are a union member**, tell your union rep what is going on. This will be in confidence, and it does not mean that a formal complaint will automatically follow. Union reps should only do what you ask them to do, and will give you the advice and support you need (see page 269). The rep could go with you to speak to the bully, or help you with a formal complaint, if it goes that far, and make sure the right steps are followed.
- **If you are not a union member**, you now have the right to *be accompanied by a companion of your choice* in any grievance hearing. This includes meetings with management on bullying (see page 271).
- **Visit your GP** See your GP if you experience the sorts of symptoms set out above. Before you go, read Chapter 21!

Keeping a diary

The following entries are taken from Teresa's diary. She worked as a secretary in a small office, with half a dozen others. Her supervisor bullied her, and expected others to follow suit. She says, 'My diary helped me to keep things in perspective and, later on, it helped me to remember certain instances. I only used the diary when things were really bad. I was often asked how did I feel at the time, and in what way was I treated differently. With my diary, I could point to the actual days.'

In the extracts from Teresa's diary in the box below, 'OM' is the office manager and 'S' is a new part-time secretary.

Wednesday 7 May: I was leaving to go home when OM announced that she had just finished a meeting which decided that the physiotherapists would hand write all their own patients' discharge letters. This removes 75 per cent of my job! Last July, it was decided that I would type them all. OM said my job will be changing yet again, but she did not say in what way... This week, S has been working extra hours. OM says my work will be less, but she is giving S *extra hours*?

Isolation

Friday 9 May: I have come through a horrible week. A and M have not uttered a word to me. S only speaks to me when she has something to say to do with work. On Tuesday morning, she made fun of me when she and P, a physiotherapist, made exaggerated comments when they said *good morning* to each other... B did say how sorry she was, that she could see how I was treated. This helps a little, knowing that someone else is aware of it.

Monday 12 May: I entered the office this morning, and although Z, M and B were in the office, only B said hello.

Tuesday 13 May: The new lady on reception will not even look at me, let alone speak to me, and I sit practically facing her all day. I don't know why this is. I have always spoken to her, but she has rarely answered me... The atmosphere in the office is very strange in that there has been a lot of whispering going on. I can't help thinking there is something up?

Wednesday 14 May: I was amazed to see S working today; she is now on four days a week. A physio told me they will not be dictating any more discharge letters from next week. This seems very odd because I do most of the letters; it's a very large part of my job. S has been allocated an extra day, yet the work is not going to be there for me to do!

Petty complaints

Thursday 15 May: I am often criticized, even if I am 30 seconds late, but S and J arrived together at 9.30 am. This is the way the office is run. It is all very worrying, because I am not accepted into the team.

Friday 16 May: They just leave piles of work in my room, with no explanation. I just think it out for myself, and do it, like adding patients' data on a system for which I have received no training.

I feel as if I am not a person. I cannot explain the way I feel, but I am sure the office manager is doing this deliberately... Certain people from medical records will not even come into the department because they say the atmosphere is so hostile... she has treated other secretaries very badly, but they leave and she just gets away with it.

16 June: [Told] I should try harder, as there were still mistakes... [they are] changing commas to full stops, and vice versa. I just feel depressed and low, and I know it is to do with the atmosphere in this department. I am criticized every day of the week until I am spinning.

Teresa says, 'It was very difficult to explain my problems to other people. How can you explain a "look", a "snigger", a "sneer" or an "attitude"? Also, how do you explain why you feel so unwell? When methods are used every working day to make you feel such a fool, it becomes hard to communicate properly with friends or colleagues.'

Teresa met her union rep and, with her support, made a formal harassment complaint against the manager. She received an apology from a senior director. But the manager escaped disciplinary action.

Bullying at work: the law

If you decide to take legal action against your employer, the options are to use employment protection legislation, or health and safety at work law. There is no separate 'anti-bullying' law in the UK, as exists for racial or sexual harassment, although bullying is often involved in both.

Bullying: the ACAS definition

There is no legal definition of bullying at work. But the independent Advisory, Conciliation and Arbitration Service (ACAS) has a definition of bullying, which is used in Employment Tribunal cases: 'Offensive, intimidating, malicious or insulting behaviour, involving an abuse or misuse of power through means intended to undermine, humiliate, denigrate or injure the recipient.'

Constructive dismissal

If you have resigned, either formally or informally, because you cannot stand the bullying or harassment for a moment longer, then you can make a claim for *unfair 'constructive' dismissal*. Your claim is made under employment protection legislation, and must be submitted to an Employment Tribunal within three months of the day you last worked.

In constructive dismissal cases, a bullied employee argues that he or she was *forced to quit* by virtue of the employer's failure to protect the employee's dignity at work. But constructive dismissal can be difficult to prove. You should be very careful about resigning if you want to be able to claim unfair dismissal afterwards.

The *key test* is whether the employer's failure to take action involved a serious or fundamental *breach of contract*. The terms of your contract can be broken in two ways:

- **Breach of an express term in your contract**. This could include, for example, an anti-harassment or anti-bullying procedure requiring your employer to carry out a proper investigation, or breaking the agreed guidelines in other ways.
- **Breach of a term implied in your contract**. Implied terms are just as powerful as the express terms, and include:
  - Your employer's *statutory duty* to take reasonable steps to protect the health, safety and welfare of employees, under the Health and Safety at Work Act 1974 (see page 24). This includes the provision of a safe place of work, and safe people to work with.
  - Common law duties, including the *duty of care* and the *duty of mutual trust and confidence*. Here, your employer has an obligation to ensure that an employee is not intimidated or humiliated, and is treated with dignity at work, and to deal with any complaints fairly and seriously.

Personal injury claims

Bullying can lead to psychological stress, depression and anxiety, as well as physical injury. Employees who suffer a work-related injury, whether psychological or physical, may be able to sue their employer in the civil courts for a breach of any of the employer's duties. (For sources of legal help and advice, see Chapter 24.)

'In imminent danger'

Some types of dismissal linked to health and safety worries are treated as 'automatically unfair', for example leaving work if you believe you are in imminent danger.

Mr McCaffrey worked as a machine minder on the nightshift with one other person, Mr Huson, a younger employee. Mr McCaffrey considered his fellow worker was behaving abusively towards him, and made a complaint. But, during the evening shift before the complaint was due to be heard, Mr Huson became abusive and threatening. Mr McCaffrey tried to telephone his manager, but Mr Huson stood right over him, shouting abuse. Mr McCaffrey walked out of the workplace, and phoned his manager and another senior employee from home. He described what had happened, and said that he would only return if the employer would take action against the harassment and guarantee his safety.

The next day, a company director spoke to Mr Huson and other employees. He took the harasser's version of events, and on this basis telephoned Mr McCaffrey and informed him that he had resigned by walking out in the middle of a shift. He sent Mr McCaffrey his P45.

Mr McCaffrey complained to an Employment Tribunal, which found that he had been unfairly dismissed. The tribunal said that the dismissal was in breach of section 100 of the Employment Rights Act 1996. This provides that it is automatically unfair to sack an employee if the reason for dismissal is that the employee left his or her workplace 'in circumstances of danger which the employee reasonably believed to be serious and imminent, and which the employee could not reasonably be expected to avert'. The tribunal confirmed that 'circumstances of danger' could include danger stemming from a fellow worker (*Harvest Press Ltd* v *McCaffrey* (1999).

Unlawful harassment

In 2003 and 2004, the Government introduced tougher Regulations dealing with harassment and discrimination at work.

If employees suffer harassment on sexual grounds, or because of their race, ethnic or national origin, religious or other belief, or for a reason that relates to their sexual orientation or disability, they can now take action at an Employment Tribunal. The Government has amended the Sex Discrimination Act 1975, the Race Relations Act 1976 and the Disability Discrimination Act 1995 with various Regulations. New rights covering age discrimination are likely to be brought in by 2006.

These legal changes are accompanied by a new general definition of harassment common to discrimination 'on grounds of race or ethnic or national origins, sexual orientation, or religion or belief, or for a reason that relates to a disabled person's disability':

> Harassment is defined as occurring when a person engages in unwanted conduct which has the purpose or effect of:
>
> (a) violating another person's dignity; or
> (b) creating an intimidating, hostile, degrading, humiliating or offensive environment for another person.
>
> The conduct is deemed to have the required effect if, having regard to all the circumstances, including in particular the perceptions of the other person, it should reasonably be considered as having that effect.

For advice on tackling bullying and unlawful harassment, you may need to contact your union rep or a local CAB or law centre (see page 263). For further information, see also the TUC's Web site: www.worksmart.org.uk.

Meanwhile, the Andrea Adams Trust and Amicus trade union are campaigning for the Government also to introduce a specific definition of 'bullying' into employment law, by supporting a Dignity at Work Bill.

Dealing with the harasser

Where an investigation reveals that there has been harassment or bullying, and that there are grounds for disciplinary action, this should be dealt with separately under the employer's disciplinary procedures.

The perpetrator should be:

- told in advance the nature of the allegation;
- given a chance to prepare and state his or her case;

- given the right to be represented; and
- dealt with fairly and objectively.

Even if the harasser is not dismissed, in deciding what penalty to apply the employer may issue a formal warning and also consider *removing the perpetrator from contact* with the person he or she has targeted.

Cath Noonan: 'I know a bit about the way they work'

Cath Noonan, a home help organizer for Liverpool City Council, received an £85,000 out-of-court settlement for a personal injury claim that followed years of sustained bullying and harassment by work colleagues. Legally, the claim was based on: 1) her employer's failure to fulfil its 'duty of care'; and 2) constructive dismissal, as she was *forced to quit* because of the employer's intolerable conduct. Cath's evidence included the findings of an earlier grievance hearing, which found in her favour; specialist medical and psychiatric evidence; and her diary, which she kept for five years. Bullies often use 'ritual humiliation' to gain power. This was true in Cath's case.

Cath started working for the council in 1986. Quite quickly, she was promoted to supervisor. But within the space of a few years she suffered the tragic loss of all three of her brothers, whose deaths from the effects of contaminated blood transfusions were used against her. The three brothers suffered from haemophilia, a blood clotting disorder. Two of them contracted HIV from contaminated transfusions of an imported blood clotting agent; the third died from hepatitis C virus.

Because she needed time off work to grieve, she confided in her manager.

> I had the misfortune to be promoted instead of a woman who had worked in the department longer than me. She was promoted later, and we worked in the same office. The first year or so was OK, but then my brothers began to die, and she really put the boot in. She would back away from me when I came into the office. But I had

> told no one but my manager about my brothers' HIV condition. Because of her getting close to the managers, I believe one of them told her.
>
> I went on sick leave and this woman watched my house. She and others she had recruited chose to spread it about that I was having work done there. Someone I know overheard her say so. She complained to managers that I was not working my proper hours. She ensured that I was excluded from the tea club, was never invited to the lunches. If I had a day off, things would be missing from my desk. I felt I was being sent to Coventry for having time off work. I spent nine months of being completely ignored.
>
> I complained to my manager, but he just told me to go and confront her. This is exactly what I had been told by his predecessor.
>
> It culminated in a manager asking me to work in a room on my own. My reply was, 'I'm not contagious'. He said, 'Well, you will have to go home'. And I didn't return again.

This was in December 1994. Supported by her union, UNISON, Cath took out a grievance against her employer for failing in its 'duty of care'. The internal investigation took 13 months to get started: initially, senior officers refused to accept the complaint, as there was no council policy on bullying. Eventually, an 18-month enquiry found in her favour. Disciplinary action was taken against one of her two harassers; nothing happened to the other. Cath took ill health retirement, brought on by work-related stress.

> It was hard enough coping with my losses, without having to confront a bully. This was social services. I kept a diary; I had heard it's good to do so. For all that time I was aware that I was being bullied, but I didn't discuss it with anyone, neither my family nor my union rep. It was only when I came home that day that I telephoned my husband and told him. I showed him the diaries I had kept for the previous four or five years. Yes, he was angry. Then we decided to involve the union.

Diagnosed by her GP as suffering from depression, Cath took extended sick leave. 'I couldn't sleep, and I had problems with my bowels. I would vomit every day before work. I was sent to the hospital for tests. They said it was because of stress, and they believed it was because of my brothers. I didn't tell the hospital about my bullying.'

She compares her experience to that of children being bullied at school. 'For every child out there being bullied, there is also an adult being bullied at work. It shouldn't happen.' Her case was handled by solicitors retained by her union, UNISON.

Further information

Bullying: The next steps, available from Mencap, the voluntary organization working with people with learning difficulties, their families and carers. Contact the learning disabilities helpline: 0808 808 1111; Web site: www.mencap.org.uk.

Stress at Work: A guide for employers, Health and Safety Executive (ref HS (G) 116), available from HSE Books (tel: 01787 881165).

Union helplines are listed on page 271. Examples of the practical advice available from union safety experts include:

- *Bullied at Work? Don't suffer in silence*, TUC leaflet, available from the TUC's Know Your Rights helpline (tel: 0870 600 4882), free.
- *Bullying at Work: How to tackle it* (1998), available from Amicus Working Environment Unit, 40 Bermondsey Street, London SE1 3UD, price £10 (non-members) or free online at: http://www.3rdsectorunion.org.uk/healthand-safety.shtml.
- *Bullying at Work: UNISON* (2000). See also *Bullying at Work: Survey report*, by Stafford University Business School. Both available from UNISON, 1 Mabledon Place, London WC1H 9AJ (tel: 020 7388 2366).
- Union advice booklets on tackling racial and sexual harassment include *Sexual Harassment*, available from Unifi (tel: 020 8879 4254), free. A list of trained bullying advisers in the banking and finance sector is available from this number.

Confidential advice

The Andrea Adams Trust. You can talk to an adviser on the helpline (tel: 01273 704900) on a completely confidential basis. The adviser will help you to clarify the issues you are facing, and suggest ways of obtaining support at work. The *Bullying Factsheet* is available from Hova House, 1 Hova Villas, Hove, East Sussex BN3 3DH, or go to: www.andreaadamstrust.org.

The Suzy Lamplugh Trust. The organization's leaflets, *Bullying in the Workplace* and *Bullying at School*, provide supportive advice and information. Available from tel: 020 8876 0305; Web site: www.suzylamplugh.org/advice/bullying.

DAWN (Dignity At Work Now), the anti-bullying support and campaign group, PO Box 11435, Birmingham B32 2WD (tel: 0774 854 0332; 01564 776748; Web site: www.dignityatworknow.org.uk).

Part 4

'New' health issues

15 'She's got PMS: that's a good one'

Toni and Louise: coping with premenstrual syndrome (PMS)

Toni works in a Midlands call centre, taking holiday bookings. She suffers from insomnia just before her period begins, during the latter half of her menstrual cycle.

> It's bound to affect how you perform at work if you haven't slept properly all night. In the first stage of my cycle I usually feel good. I look forward to going to work. But, during the second half of my cycle, I often need to withdraw for a couple of days. I need quiet, no stress put on me. Sometimes, if I can just be quiet for a while, I will be fine. Yet I also want to be comforted, whether I am at home or at work.

Louise is a specialist adviser on a dot.com help desk. As with many women suffering from acute PMS, the symptoms have increased in severity with age. 'Some people are able to control it, to some degree, but mine has become so severe, especially over the past few years, that I really haven't been able to do so. I handled it by having to come clean with my employers. So, to some extent, there has been a certain level of understanding at work.'

An ovarian tumour aggravated her PMS symptoms, to the point where the changes to her behaviour were so obvious that she had to explain to her supervisor what was going on. 'He was perhaps a little unsympathetic at first, but he now has more understanding of the condition. He takes it on board more, understands that I have bad days and good days. I have also informed my department manager. They haven't been that bad at work, really. But it has taken a long time to achieve recognition.'

Less trust, more pressure

Louise is the only woman in her team of seven advisers.

> They know about my PMS, and I suppose I take a lot of flak about it in some ways, in terms of jokes and things like that. Some people in the office have been more understanding than others. Being honest can make you more vulnerable. You have to allow for a side to some people that shows itself in these circumstances. As the saying goes, they will kick you when you are down. If you come out and say you have PMS, you are exposing yourself even more. Particularly now, when we are all under pressure at work.
>
> I know I can be wound up when I am in that vulnerable mood. If they know, and they can see it, they can provoke me, and I will respond. Naturally, you can get yourself into trouble for being like that, if the reason is not fully understood. But I think PMS exacerbates the situation. It's hard to go public because not all the people you work with will be receptive to it.

Louise argues that new working practices have brought a new, tougher culture. Every task in the call centre is monitored and measured down to the finest detail. The pressure of calls can be very high, and PMS is aggravated by stress. 'If you have PMS symptoms, they are likely to be exacerbated as you approach the menopause. They have for me, and the fact that the working environment is less tolerant, less flexible and more closely monitored has made my situation more difficult.'

Coping through control

> The best way to deal with stress is to be able to exert some influence over your working environment, and the way you do things. While you cannot expect to do this in every situation, there are some ways to achieve this.
>
> My supervisor used to make me go on the phones on those days when I really was not up to it. I was not in the right frame of mind at all. Now, I can undertake other tasks that are just as central to my job. The next time I'm rostered on the phones, I'll work as hard as the next person, if not harder. And having this flexibility and understanding makes me more committed to my employer.

> I'm not saying I need three days off the phones each week, or anything like that, but there are occasions, although each month is different, when I need that flexibility. There is some work I can do perfectly well most days, which I can't do on others. It is all about having a little bit more flexibility in the job, while still being able to make a contribution.
>
> Until recently, you were more trusted to do your work. I am not a lazy person, and I never have been. But sometimes you need a little more understanding because, for maybe three days each month, you may only feel up to doing certain tasks. There's definitely a cycle of attention and inattention at work related to PMS. However, the contribution you make at other times is every bit as effective.

Louise recognizes an employer might interpret these demands as coming from someone who was not doing the job properly. 'It's hard for employers to understand the difference, because employers appear to know so little about PMS. That's half the problem. It is an awareness thing. Through time, the people I work with have heard more and more about the condition, and can see the cyclical nature of it.'

Pull yourself together

Toni suffered from severe postnatal depression at the age of 21. 'I learnt a long time ago about the old "Pull yourself together" bit, the "Oh, what a load of old rubbish that is", the looks that say she's a bit off her trolley, or "She's got PMS; that's a very good one; it's just an excuse".' She realized:

> You can't just tell anyone about this sort of thing. Even though the level of awareness is slowly changing, when I filled in my medical forms for my current job, I didn't declare my PMS symptoms, in case I was discriminated against.
>
> That is the thing about PMS. We can find it very hard to deal with the kind of response that says you are just using it as an excuse. Sometimes the onset of the symptoms can arrive without you realizing. It's not until someone says to you, 'Are you all right? You don't seem like yourself today' or 'Wha-hey!' and step back a pace at some extremely blunt response of mine, that's when I realize.
>
> I am patient with my customers – most of the time! – so perhaps this comes out more with colleagues. Sometimes, talking to the girls at work, I will forget what I am saying in mid-sentence. I've totally gone.

She says moderate to severe PMS is a kind of 'Jekyll and Hyde' syndrome. 'You do change, and I do honestly think that anyone with any grain of sense, even if they don't suffer from it themselves, cannot help but notice the change in someone else's behaviour. It's even in your facial expressions.'

The importance of good diet

Louise has learnt the hard way the importance of eating regularly. She would not eat much at all for the first two weeks, and then binge in the second half. Restrictions on eating at work did not help.

> Instead of being able to have a little something to eat at my desk when I needed it, I had to wait until my allotted break time. My blood sugar level was down on the floor, and I was shaking and hypoglycaemic, because my blood sugar level had gone. I couldn't concentrate. You don't know what you are talking about. Management wouldn't treat a diabetic at work like that!

Timing

> If you were to monitor my work performance you would probably find that it was about normal in the first two weeks of my cycle, but below average in the second two. You would find that, if I were to make mistakes and to forget things, you could relate it to the second half of my cycle.
>
> I try to relate my PMS cycle to my job. I am taking calls every working day, dealing with the public, never knowing what's on the end of a call when you answer it. I know where I am because I take Cyclogest [a progesterone pessary] every night between the 12th and 26th days of my cycle. I am aware as I go through my cycle of my body's different responses. I start with some pelvic pain; then I move on to the bloating as I'm due to ovulate. I then begin the various other symptoms, which are exacerbated by work if it is very stressful.

Coping routines

Toni finds that her symptoms are much more severe if they coincide with a lot of pressure at work. She suffers from hormone-related insomnia, so that she does not sleep properly as she is approaching her period.

> I get mood swings. I can't concentrate or retain information. I cope by completing all my paperwork straight after every completed call. I finish off on screen any special requests for customers. A lot of people in the office don't do this, but I have to do it there and then. I can't wait until a slack period, because I know that if I do, I will forget what's needed.

Although it eliminates mistakes, this routine can slow down the pace of her work, and she may be taking fewer calls at this time than her fellow workers. So far, no one has mentioned this to her, although she expects her manager to comment at some point.

'I guess we are talking about work culture. PMS is exacerbated by stress. We get exaggerated responses, exaggerated emotions. You can get really upset about something that, at another time in the month, would never bother us. I think to myself, you know this is stupid. It can be scary sometimes.'

Nevertheless, she feels that she is a trusted member of the group sales team. 'One of the positive things about company culture where I now work is that I don't have people breathing down my neck all day.'

Relationships

> We were extremely busy at work recently, coinciding with the beginning of my cycle. I warned my kids and my partner that it is going to be worse this month. "I am telling you now, because I have had so much stress in the first half of my cycle. When I get into the premenstrual stage, I am going to be worse." And sure enough, I was. Not because I conditioned myself to believe it, but simply because that is what happens.
>
> My partner is very tolerant. But PMS does affect all of your relationships, at work or home. Everybody has rows, but PMS is stressful. The things you can say when you are like that! I am embarrassed afterwards, because you are just angry. The worst thing is, there are times when you would like to be left alone. People can go too far with you sometimes.

16 PMS and the menopause

Both PMS (premenstrual syndrome) and the menopause raise health and welfare issues for hundreds of thousands of women workers. But neither health issue is yet taken seriously by most UK employers.

EastEnders actress Race Davies played the part of PMS sufferer Jackie Owen, in a major storyline that ran over summer 2000. Viewers witnessed the confusion and trauma of Jackie and her partner, Gianni, as they struggled to come to terms with her severe PMS. Never before had the effects of PMS been so graphically illustrated in the media.

By definition, PMS affects working women. But Chris Ryan, head of the National Association for Premenstrual Syndrome (NAPS), says that to the best of his knowledge there are no 'good practice agreements' for PMS in the workplace. Similarly, in 2003, a TUC study showed that only 2 per cent of employers' health and safety policies cover issues related to the menopause.

This chapter examines:

- how both PMS and the menopause affect working women;
- what employers and unions should be doing;
- practical advice to help you cope at work.

Explaining PMS

NAPS defines PMS as: 'The cyclical recurrence, in the latter half of the menstrual cycle, of a combination of distressing physical, psychological and behaviour changes. In severe cases, these changes can lead to a breakdown of relationships at home, at work and among friends. Your normal lifestyle can be severely disrupted'.

Dr Nick Panay, consultant gynaecologist at The Hammersmith Hospital, estimates that up to 90 per cent of women experience some changes premenstrually. However, between 5 and 10 per cent of women – about 800,000 women in the UK – are so severely incapacitated that it dominates their life during this phase of the cycle. A further one-third of women experience 'moderate' physical discomfort or psychological ill health.

There are about 16 million women in the UK of menstrual age (ie 12 to 51 years of age). Of the 800,000 women with severe PMS, two-thirds, or nearly 600,000, are working full or part time.

Dr Panay identifies around 160 individual symptoms – physical, psychological or behaviour-related (see page 194 for examples). Most women have their own characteristic combination of symptoms (or 'syndrome'). 'Even within the same person, the severity of these symptoms often varies from month to month,' he suggests.

Winning recognition of PMS

NAPS recognizes that the very existence of PMS is still disputed. The organization's autumn 2000 national conference heard many instances of GPs still not believing that there is such a condition, and consequently failing to treat it.

NAPS surveyed nearly 500 women who had called the NAPS helpline, having already visited their GP. Of these, 42 per cent stated that their GP was either unsympathetic or did not seem to know much about PMS. There are still only 15 specialist PMS clinics in the UK. You have to go through your GP for a referral to a specialist clinic. In NAPS' view, if your GP is not sympathetic, change your GP.

NAPS says, 'PMS is a 20th century phenomenon, reflecting changes in social structure and lifestyle expectations'. In the past, there were fewer opportunities for the full effects of PMS to become apparent. The onset of puberty was a much later occurrence. For many women, the years between puberty and menopause were likely to be filled with several more pregnancies, and the conditions offset by longer phases of breastfeeding. Nowadays, with the average age of puberty at 12 years and menopause at 51 years, together with contraception reducing the number of pregnancy-related pauses in the monthly cycle, a woman may expect to experience in the order of 470 menstrual cycles during her life. Hence, the effects of the menstrual cycle now tend to be far more dominant in a woman's working and domestic life.

Symptoms appear at some time during the second half, or 'luteal phase', of the menstrual cycle. This is the time between ovulation and menstruation. Symptoms then disappear, or at least significantly improve, either on the first day of the period, or after the day when the flow is heaviest.

Dr Panay says, 'There is no single theory to say what brings on PMS. But the function of the ovaries is the key to understanding the condition, particularly the *rapidly changing levels* of the hormones oestrogen and progesterone during the monthly cycle' (see Figure 16.1). After menstruation, you should then be symptom-free for at least a few days. Women who experience symptoms during these few days are less likely to be suffering

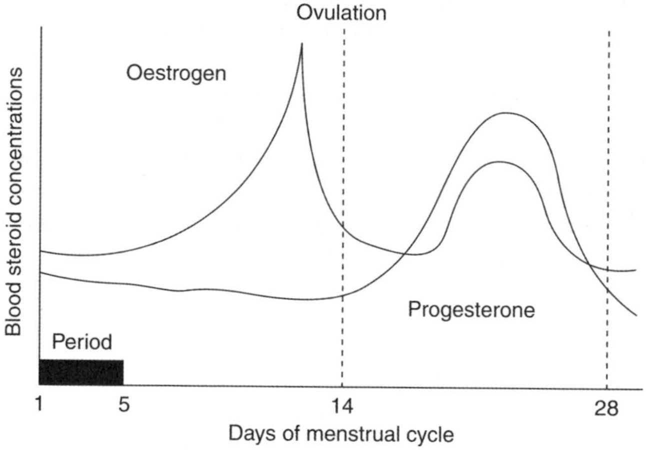

Figure 16.1 Hormone levels during the monthly cycle
(Source: *Understanding PMS: Help and support for PMS sufferers and their families*, NAPS)

from PMS, and should visit their GP for an opinion. However, many women get a spell when symptoms return just after their period.

Certain events are linked to the onset of PMS, or can increase its severity:

- hormonal events, such as childbirth, ceasing to use an oral contraceptive, or the time leading up to the menopause;
- increasing age; and
- stress.

'There is no doubt that PMS interacts with many aspects of life, especially at difficult or stressful times,' he comments.

Self-diagnosis

There are no laboratory tests that make PMS identification simple. Correct diagnosis of symptoms is an essential first step for successful therapy and treatment. A successful diagnosis can be made by keeping a 'symptoms diary' over a period of at least three months. A personal diary is a far more reliable method than trying to remember symptoms. It provides written evidence of the precise date of menstruation and the symptoms experienced.

On a daily calendar, a symbol is used to identify the two or three worst symptoms, for example M for menstruation, H for headache, I for irritability, B for breast tenderness. The severity of each condition can be scored on a scale of 0 to 3. Charts are available from NAPS, together with a personal symptoms checklist.

Four steps to managing PMS

First step: visit your GP

In general, the best treatment for women with PMS is a sympathetic GP and practice nurse. The most appropriate therapy takes account of the woman's own assessment of the type and severity of her symptoms. This is why the diary is so important.

Reassurance that she is not 'going mad', and that PMS affects many women to a greater or lesser extent, can itself be very therapeutic. All sufferers benefit from simple advice related to dietary changes, exercise, relaxation, stress avoidance and lifestyle modification. 'The psychological benefits of discussion, counselling and education cannot be overemphasized,' Dr Panay suggests. Not all women require medical help. Symptoms can be sufficiently controlled by changes to diet and lifestyle, and stress management, especially at work (see below). And, if one option proves to be ineffective after four monthly cycles, another may prove to be more suitable.

Second step: dietary advice

Many PMS sufferers report the beneficial effects of simple dietary changes and better eating habits. Gaynor Bussell, consultant nutritionist and dietician, is dietary adviser to NAPS. In her booklet, *Dietary Guidelines for PMS*, she strongly advises regular, healthy eating – see guide, page 195. This includes a healthy, low fat, high fibre diet, and avoiding gaps of more than about three hours without eating.

Third step: coping at work

There are no hard-and-fast rules to follow in coping with PMS at work. We are not aware that any leading employer has established a 'good practice' PMS policy, although one or two unions have published guides.

The following advice is based on NAPS booklet, *Understanding PMS*, and the first-hand experience of PMS sufferers we have interviewed:

- **Eat regularly**. Follow the *Dietary Guidelines* suggested here. Start the day with a good breakfast. Try to avoid long spells at work without eating, especially during the luteal phase, when PMS is present. Take snacks to work to eat during the morning and afternoon. Spread your intake of energy-giving foods so that the blood sugar level doesn't fall too far at any point during the working day. These foods might include a sandwich with a low-fat filling, fruit, plain biscuits, scones, buns. If you have a long or late meeting, try to eat something beforehand. Taking 5 or 10 minutes away from your job in mid-shift to eat may be easier said than done, and may need permission from a supervisor.

- **Drinks.** Drink less caffeine, tea and cola (see box, page 195). Substitute more water, herbal teas and decaffeinated or non-carbonated drinks.
- **Telling your fellow workers?** Louise's experience is that telling people you work with about your symptoms is a difficult step to take. But she adds, 'If you don't, variations in your contribution at work can easily be misconstrued. Talking to people you can trust may help you to cope with work stresses in the most difficult days. But be careful: you are exposing yourself as well. It can also make you more vulnerable.'
- **Informing management.** A line manager, personnel officer or occupational health nurse who understands your needs is more likely to accept variations in the pace or type of work you undertake, for one or a few days each month. *Flexibility* is surely a better option for the employer than the alternatives of you working less well under stress, or taking whole days off altogether.

Tips for tackling stress

'Stressful situations make PMS worse.' This is the blunt conclusion in *Understanding PMS*. These suggestions may help:

- **Try to gain more control over your job**. Toni learnt to *pace herself* at work. She takes telephone bookings for holidays, and, during the most difficult days of the cycle, she deals with all the paperwork at the time of the call in order not to forget any important detail. At other times, she will move straight on to the next call, dealing with paperwork later. The bottom line is that her work remains accurate. Louise emphasizes the value of *team support*. There are always opportunities to return that support. A supervisor might object to making exceptions for some staff. But they are only adjusting work for a day or two, as necessary. Expecting a negative response from supervisors, a lot of people will take the whole day off work. It's about employers being *adaptable* to the needs of their staff.

- **Plan ahead; be prepared**. You will know when your PMS symptoms are likely to be more severe. So, when you have a job interview or a performance, try to rearrange it to avoid the week before your period. The best time would be the week after the period.
- **Relaxation**. There will be times when you cannot rearrange an interview, test or appointment. So it is important to learn relaxation techniques. The ability to relax in stressful situations will be useful, not only in managing your PMS at this time, but also in calming you down. The relaxation you choose could be a quiet hour to yourself, reading a book or having a long soak in a hot bath, or yoga. Your GP may be able to advise on relaxation exercises.

Fourth step: clinical treatments for PMS

A wide range of clinical treatment options is available to women suffering from severe PMS, for whom dietary and lifestyle changes do not deliver sufficient relief or control. They range from the use of evening primrose oil for breast tenderness, to oestrogen patches or implants, which can help to suppress the ovarian cycle. However, detailed clinical advice is beyond the scope of this book. Seek further advice from your GP, a specialist clinic and NAPS. (See also *Premenstrual Syndrome: A clinical review*, Further information, page 197.)

New workplace policies for PMS

NAPS is clear that employers need to develop workplace guidelines for PMS sufferers. They could start by *nominating a designated person* to provide advice and support. Ideally, every employer and union should do so. Personnel departments could *raise awareness* by posting PMS advice with other employment information on company intranets. This could cover PMS symptoms, when they are likely to occur and how they might affect work, together with suggested ways to cope and how to adapt work to the needs of the employee when it becomes necessary.

Employers should review their 'absence management' procedures. PMS-related absences from work are likely to show up as regular days off 'sick'. Return-to-work interviews with staff may be targeted on regular absentees, and may be more concerned with stopping than with understanding the absences. Health policies should be reviewed to recognize the prevalence of PMS among short-term absences, and encourage flexible responses to it.

Some unions, such as USDAW, provide basic healthcare guides for PMS sufferers, advising women to contact their local union rep for advice. Developing recognition of PMS among UK employers and unions is a high priority for NAPS.

Men and PMS

How men support partners, wives, work colleagues or friends can be crucial in helping women to come to terms with, and to control, the symptoms. As one of the delegates to the NAPS conference remarked, 'So much of PMS suffering is kept within the four walls of the home'.

George and Mike (not their real names) are volunteers on the NAPS helpline. Mike says a common question from male partners who telephone the helpline is, 'Why is my wife so unwell?' Another is, 'She's got medication; why isn't it working?' He will help the caller to identify the main symptoms and their pattern over a monthly cycle. He will explain that there isn't one single treatment.

George says:

> Men find it very difficult to understand that there can be this condition which may be beyond their wife's control. When I came in the back door, I never knew whether I was going to eat dinner or wear it. And because it happens at home, the man thinks, 'It must be me'. They were never taught about it at school, except, 'Look she's got a period; what a giggle'.
>
> I was lucky in my company. I advised them about my wife's condition, so I can phone home, for that bit of support when she is feeling bad. For the sake of a few seconds on the phone, just to give

reassurance, can save employers a lot of time off work later on for something more serious.

The helpline advises the male partner to give support, encourage his wife or partner to eat regularly, talk about symptoms and ways to help her avoid stress.

PMS symptoms

Physical symptoms, according to NAPS, include:

- bloating;
- weight gain;
- headaches;
- skin rashes;
- breast tenderness;
- pelvic pain;
- constipation or diarrhoea;
- puffiness of face, abdomen or fingers;
- muscle or joint stiffness, and backache;
- abdominal pains or cramps;
- exaggeration of conditions such as migraines, asthma, epilepsy.

Psychological and behaviour symptoms include:

- irritability;
- aggressiveness;
- tearfulness and 'feeling low';
- poor concentration;
- anxiety, tension or unease;
- insomnia;
- appetite changes;
- mood swings and depression;
- altered interest in sex;
- food cravings;
- loss of self-control.

Dietary guidelines

The right diet can bring PMS relief:

- **Feel-good factor**. A well-balanced diet that is low in fat and high in starchy foods helps the production of a 'chemical messenger' in the brain called serotonin which, when released, can help lift mood and boost the 'feel-good factor'. To manufacture serotonin, the brain needs certain vitamins and minerals, particularly those derived from wholegrain cereals, fruit and vegetables, and a frequent supply of carbohydrates.
- **Breast sensitivity**. A hormone-like compound, prostaglandin E1 (PGE1), is believed to be in short supply in PMS sufferers. When PGE1 levels are low, the body may become ultra-sensitive to the changes in hormone levels that occur before a period. This gives rise to such symptoms as breast tenderness. The body's ability to manufacture PGE1 is impaired by excess consumption of alcohol, caffeine and dietary fats.
- **Caffeine**. Too much caffeine can hinder the absorption of vitamins and minerals, so try not to drink more than five cups/glasses of tea, coffee and cola a day. Substitute with diet carbonated drinks, herbal teas, decaffeinated tea and coffee, and water.
- **Alcohol**. For women, the recommended maximum weekly alcohol intake is 21 units. Experts recommend that PMS sufferers should not exceed 14 units. One unit is: a half-pint of lager or beer, one small glass of wine or one measure of spirits. Alcohol can suddenly lower the blood sugar level, and affect sleep and tension.

Source: *Dietary Guidelines for PMS*

Working through the menopause

Working through the Change, a March 2003 TUC report, shows that seven out of 10 women aged 45 to 59 years are in work. Most women go through the menopause, or 'the change', between the ages of 48 and 55 years. But only one employer in five even provides information about the menopause, the TUC says.

Managers are often critical of menopause-related sick leave, while women find it embarrassing or difficult to mention it to their supervisor. According to the TUC study, menopause symptoms that are likely to make work worse include:

- hot flushes;
- headaches;
- tiredness;
- anxiety attacks; and
- short-term memory loss.

Some job hazards make matters much worse. Working in a stressful environment tops the list, followed by the discomfort of working in a fixed position. But manual handling and any kind of hot work, such as working in a kitchen, can also present difficulties. Managing or supporting others, and dealing with members of the public, can also be very hard to deal with.

The TUC report *Working through the Change* recommends that employers should:

- provide cold drinking water, and easily adjustable temperature and humidity controls;
- encourage flexible working and reduce long hours;
- avoid penalizing staff for taking frequent toilet breaks;
- provide quiet rest facilities and advice for female employees;
- avoid negative and discriminatory attitudes towards older women at work by recognizing the potential problems related to the menopause;
- provide appropriate information and training to their managers;
- develop policies in consultation with unions on the menopause to cover sickness absence, paid leave for treatment, occupational health screening, flexible working patterns and rest breaks.

The TUC also wants the Health and Safety Commission to produce guidance for employers and employees on the menopause. And the TUC is calling on unions to provide women workers with advice about the menopause and employment rights, and to develop awareness-raising courses for union reps. See also the TUC Web site: www.tuc.org.uk/menopause.

Further information

Menopause, Hazards Factsheet 82, www.hazards.org.uk.

National Association for Premenstrual Syndrome (NAPS), 41 Old Road, East Peckham, Kent (tel: 0870 777 2187; helpline: 0870 777 2177; Web site: www.pms.org.uk). NAPS booklets include:

- Panay, Dr Nick, *Premenstrual Syndrome: A clinical review.*
- Bussell, Gaynor, *Dietary Guidelines for PMS.*
- *Understanding PMS: Help and support for PMS sufferers and their families.*
- *NAPS News*, quarterly newsletter.

Premenstrual Syndrome and Period Pains and *Menopause*, available from USDAW (tel: 0161 224 2804; Web site (to download leaflets): www.usdaw.org.uk/site_documents).

The Menopause Exchange, regular newsletter and source of practical advice, PO Box 205, Bushey, Herts WD23 1ZS (tel: 020 8420 7245, e-mail: mexchange@btinternet.com).

17 Driven to death

The couple were driving from their home in Staffordshire to Edinburgh when they joined a traffic queue to join the motorway. Behind them, a young HGV driver took a call on his mobile phone from his employer. He was late for the next delivery. A few moments later, the lorry collided with the car, and its passenger was killed.

Road traffic accidents are the biggest cause of sudden death at work.

More people than ever are driving as part of their duties. The TUC estimates that around 300 of the 1,200 drivers killed on Britain's roads every year are at work.

Far more people die at the wheel of their company car, van or lorry than at their place of work. In 2002/03, there were 226 fatalities at work, of whom 182 were employees and 44 were self-employed.

In response to pressure from the TUC and the Royal Society for the Prevention of Accidents (RoSPA), the Government set up a task group on work-related road safety. Its report in spring 2002 called on the Government to make wide-ranging changes to improve driver safety. Some of these proposals have been taken up and we look at them here: new duties on employers to eliminate risks to drivers; and a ban on the use of hand-held mobile phones. But many other proposals have not yet been taken on board, like changing police report forms so that they record the purpose of the journey, or requiring employers to report fatal and major *road accidents* under the RIDDOR regulations (see page 8).

A dangerous occupation

According to RoSPA, driving long distances (around 25,000 miles a year) is the third most dangerous occupation in the UK. This amount of driving is almost as risky as working down a coal mine (see Table 17.1). There is a higher chance of being killed at the wheel of a car than being killed working on a building site. RoSPA estimates, based on official accident-at-work statistics, show that anyone habitually driving 25,000 miles a year stands a 1:8,000 chance of being killed.

The TUC estimate that 300 people are killed on the roads each year while driving as part of their job is based on official traffic survey data. These show that about a *quarter* of all the mileage driven on roads in Britain is work-related. Using figures for 1998, the TUC showed that 3,500 people of all ages were killed on the roads. Of these fatalities, one-third, or 1,200 incidents, involved people driving a car, van, lorry, bike or motorcycle. The TUC assumes that a quarter of these drivers, or 300 people, were driving as part of their job.

The police currently record the *type* of vehicle involved in fatal crashes, so we know the annual toll of drivers of light goods and heavy goods vehicles killed on the roads. In 1998, 46 drivers of lights goods vehicles and 52 HGV drivers were killed. These figures demonstrate that the majority of work-related deaths at the wheel involve a car driver.

Table 17.1 Annual average probability of a work-related fatal accident

Occupation	Risk of Death
Deep sea fishing	1 in 750
Coal mining	1 in 7,100
Car driving: 25,000 miles a year	1 in 8,000
Construction	1 in 10,000
Agriculture	1 in 13,500
Service industries	1 in 150,000

(Source: Royal Society for the Prevention of Accidents)

Tackling road risk

The independent Work-Related Road Safety Task Group accepted that at least one-third of all serious or fatal road accidents involve a person who was at work at the time. In response, the HSE issued new Guidance for employers, *Driving at Work: Managing work-related road safety*. It says that 'road traffic accidents may account for over 20 fatalities and 250 serious injuries every week'.

Some employers believe, incorrectly, that, provided they comply with road traffic law requirements, such as having a valid MOT certificate, and their drivers have a valid licence, this is enough to ensure the safety of their employees when they are on the road. But this is no longer the case. Published in September 2003, the HSE's Guidance shifts responsibility for driver safety squarely on to the shoulders of employers.

The HSE's Guidance applies to any employer, manager or supervisor with staff who drive, ride a motorbike or ride a cycle at work, and it particularly applies to fleet managers. It also applies to self-employed people. But employees and union reps will also find it useful. Many will find it groundbreaking advice, requiring a whole new set of safety policies and attitudes.

Driving at Work informs employers that they have a duty under the Management of Health and Safety At Work Regulations (see page 35) to carry out an assessment of the risks to health and safety faced by their employees. Risk assessment now extends to at-work driving. And, as the HSE points out, 'promoting sound health and safety driving practices and a good safety culture at work may well spill over into private driving'.

The main issues employers have to deal with are:

- **Policy**: ensure that the health and safety policy covers work-related road safety.
- **Responsibility**: establish top-level commitment to road safety in the organization. Does the person have enough authority to exert influence if changes are needed?
- **Safety systems**: ensure that the organization assesses risks and operates a procedure to manage work-related road safety effectively.

The HSE's Guidance sets out the same basic five-step approach to driving as the risk assessment for any other work activity: identify the hazards involved; decide who might be harmed (usually the driver); evaluate the risks and decide if action is needed; record the findings; and review the risk assessment from time to time.

There are still at least two significant gaps in the HSE's approach to work-related road safety: 1) It seems that the police will not be collecting information on the purpose of the journey when road traffic accidents are investigated until 2005. 2) Although the task group wanted work-related driving accidents to be reportable under the RIDDOR Regulations, there is no date for this change to take place. At present, health and safety inspectors only investigate road accidents away from the workspace when specific work activities are involved, such as refuse collection. The HSE says it will hold a public discussion on the issue in 2004. When these gaps are filled, then safety inspectors will be expected to work with the police to consider whether management failures have contributed to the accident or loss of life.

Road risk assessment

RoSPA has played a key role in encouraging employers to take a 'risk management' approach to occupational road risk. Controlling road risks should become part of an employer's 'safety culture'.

The RoSPA approach is:

- **Step 1: Audit vehicle usage, accidents and their costs**:
  - the number of vans, lorries and essential or casual car users;
  - journeys (current mileage by types of vehicle, length of journeys);
  - date, time, place of accidents, severity of injuries;
  - annual cost of insurance, repairs and absences from work following road accidents.
- **Step 2: Carry out a risk assessment**. In occupational safety, a hazard is something that has the potential to cause harm. Risk is the likelihood that that harm will arise from a hazard (see

page 29). RoSPA identifies the factors likely to increase the likelihood of an accident, and recommends using a scoring system to highlight issues of highest concern. Table 17.2 is adapted from the RoSPA report, *Managing Occupational Road Risk*.

- **Step 3: Introduce control measures**. RoSPA recommends the following checklist of control measures for employers intent on managing road risks. But many of these ideas are also sound advice for the individual driver:
  - Eliminate unnecessary journeys: check alternative means of communication.
  - Change mode of transport, eg train / drive or fly / drive.
  - Avoid driving in adverse conditions: at night, in poor weather.

Table 17.2 Scoring the road risks

	Score		
Factors to Consider	**Low**	**Medium**	**High**
The journey			
Road type			
Overall distance			
Night or day			
Weather conditions			
Adequate / inadequate breaks			
Adequate / inadequate time			
The vehicle			
Maintenance standard			
In-car distractions			
Crash resistance			
The driver			
Age			
Experience			
Driving competence			
Stress			
Fatigue			
Accident history			
Attitude			

(Source: adapted from RoSPA)

- Reduce distances: set maximum driving distances per day, per week, per year.
- Control drivers' hours: set upper limits for driving hours each day, week, month.
- Specify safest routes.
- Specify safer vehicles.
- Set driver capability standards, eg through requiring drivers to pass an advanced driving test, paid for by the employer.
- Require regular eyesight tests: the legal visual standard for a car or motorbike driver is to be able to read a number plate at 20.5 metres, with glasses if worn.
- Set clear policies on alcohol and other substances.
- Prohibit the use of mobile phones, including hands-free sets, whilst on the move.

Brake, the national road safety campaign, publishes a checklist to help drivers audit their employer's road safety management. It covers safety culture and priority given to road risks, driver safety and training, and vehicle safety.

Mobile phones

Driving while using a hand-held mobile phone is a specific offence under Road Traffic Regulations. From 1 December 2003, offenders will initially be subject to a £30 fine, which can be increased to a maximum fine of £1,000 if the matter goes to court. The Government is also planning to legislate to make it an endorsable offence, so that drivers will get three points on their licence each time they are caught holding a phone.

Research has demonstrated that if you drive and use a mobile phone you are four times more likely to have an accident. Hands-free calls are also distracting and drivers should be aware that they still risk prosecution for failing to have proper control of their vehicle, for careless or even reckless driving if use of a phone affects their driving in this way.

RoSPA has long campaigned for a complete ban on the use of mobile phones by drivers when they are driving, and employment protection rights for employees who don't take a call from their employer whilst at the wheel. Its research shows that using a mobile phone while driving greatly increases the risk of having an accident. The risk applies whether the phone is hand-held or hands-free. The key is the loss of concentration. The danger lingers on in the minutes after the call has finished.

Some employers, like Permabond, already prohibit their drivers from using mobile phones, including hands-free sets, whilst driving. Civil service union IPMS advises against using any type of phone while driving.

The Government's safety leaflet *Mobile Phones and Driving* says, 'You are not in full control of your vehicle if you are holding a mobile phone while driving. Keep the phone switched off. Only use the phone after you have stopped in a safe place.'

'Nod off and die'

Falling asleep at the wheel accounts for one-fifth of all accidents on non-urban roads, according to Professor Jim Horne of the Sleep Research Laboratory at the University of Loughborough. These accidents are worse because of the high speed of impact. Sleep kills at least as many people as alcohol. He says, 'It is quite common for workers to be driving around with a similar level of impairment as someone at or beyond the legal limit for alcohol. Yet there is no similar level of public disapproval for those driving around sleepily. But these people are twice as likely to kill someone as in a normal road shunt.'

His research is based on laboratory tests using vehicle simulators, plus new analysis of road accident data. Characteristics of sleepy accidents include:

- They occur on dull roads, usually early in the morning, at times when your body clock signals sleep: between 4 am and 6 am, especially the 'early start to beat the rush'.
- They involve speeding and driving too close.
- There is no evidence of braking before the accident.
- There are no personal medical defects.

- The driver could see the impact point several seconds before the collision, ie there is prolonged inattention.

Nightworkers and truck drivers are especially at risk.

Questioned by the police, drivers deny falling asleep for fear of prosecution and loss of insurance cover. Many also suffer 'traumatic amnesia', and cannot recall the collision. Drivers do things to keep themselves awake, such as shifting around in the seat, opening the window or gripping the wheel hard. Professor Horne describes these as 'danger signs'. Yet, just as we have a poor ability to recall the feeling of hunger or thirst, we also have a poor recollection of feeling sleepiness. So, although drivers don't remember feeling sleepy before an accident, at the time they *are*, of course, aware that they feel tired. When fighting sleep, you must recognize that you are already sleepy. Do something about it.

What you can do for yourself

The sleep research centre tested a wide range of counter-measures to tackle sleepiness. By far the most effective formula is **to stop and drink two cups of caffeinated coffee, followed by a 15-minute nap**. The caffeine kicks in after 15 minutes. Laboratory tests showed this approach cuts driver accidents by 91 per cent. The caffeine content of each cup of coffee varies widely, from 35 to 80 milligrams. The strength of taste is not related to caffeine content. This is why you should take two cups. Energy drinks, available at service stations, can have 75 milligrams of caffeine in a standard dose. Dosing up with caffeine is a one-off measure. Do it once if necessary; it is not to be repeated. The next time you feel tired, take a proper break; stop the journey.

Other counter-measures, such as cold air, taking exercise and turning on the radio, were all found in lab tests to be 'practically useless'.

Personal safety

Harassment or actual attacks on drivers are relatively rare, but they do happen. As the Suzy Lamplugh Trust points out,

'Forethought can give you more confidence and minimize the risk.' The trust publishes practical personal safety guidance on:

- driving, including route planning, being forced to stop, motorway breakdowns;
- car parks: where to park, basic safety steps;
- road rage: keeping cool, avoiding eye contact, dealing with 'car stalking'.

Further information

Driving at Work: Managing work-related road safety, from HSE Books (tel: 01787 881165; Web site: hse.gov.uk/pubns/indg382.pdf), free.

Guidelines on Safer Driving in the Workplace, available from IPMS trade union (tel: 020 7902 6600).

Managing Occupational Road Risk: The RoSPA guide, Royal Society for the Prevention of Accidents (tel: 0121 248 2222; Web site: www.rospa.org.uk), price £20.

Mobile phones and Driving, www.think.dft.gov.uk/mobile/index.htm.

Road Safety Management: Self-audit, available from Brake (tel: 01484 559909).

The Suzy Lamplugh Trust, www.suzylamplugh.org.uk.

18 Teleworking

Teleworking – working from home – is growing rapidly. More than 2 million people work from home for at least a day a week, using a computer and a telephone link to their employer, or to clients. For some people, teleworking provides an answer to fundamental problems, a way to strike a better work–life balance and cut the daily cost, stress and inconvenience of commuting. Meanwhile, for many people with disabilities, teleworking opportunities help overcome the barriers of not being able to commute to an office.

There are two categories of homeworkers: those who work *at home*, such as teleworkers and traditional outworkers; and those who work *from home*, such as field sales reps or mobile workers.

As with any other working arrangement, telework also brings particular health and safety risks. If you work at home, your employer still has health and safety duties towards you, including providing a safe place of work and safe working practices.

If you already work at, or from, home you can check your current working arrangements against the practical advice offered here, such as safe use of DSE equipment, or personal safety.

If you are contemplating teleworking, then you have the opportunity to take a systematic approach to this new way of working.

Setting up to telework

One of the key problems associated with teleworking is the danger that, once you are out of sight, you may not be subject to the same level of protection and supervision as someone based in a traditional office.

Practical considerations include:

- **Ground rules**. There may be certain ground rules about separating domestic and work tasks; your availability if contacted; your workload; and reporting requirements.
- **Space**. Do you have the minimum necessary space at home? The area to be used as workspace should be agreed with your employer beforehand.
- **Home safety assessment**. Your employer should visit your home to safety-assess your working arrangements.
- **Teleworking agreement**. Your organization may have a teleworking agreement. Obtain a copy; check your entitlements. Discuss the arrangements with your union safety representative.

But, as the *Teleworking Handbook* points out, only two-thirds of workers received health and safety advice when they started work at home. Just under half received no advice at all, or assistance or cash support. For one in three teleworkers, the 'office' is part of a living room, or the kitchen table.

Working safely at home

Your employer has a duty to ensure that your homeworking arrangements are safe, and comply with the Display Screen Equipment (DSE) Regulations 1992 (see page 36). These Regulations are summarized in a Health and Safety Executive (HSE) booklet called *Homeworking: Guidance for employers and employees*, available on the HSE's Web site.

The HSE guide sets out a basic five-step approach to risk assessment for homeworking. Employers should either send a competent person to your home to undertake the assessment or, if this is not practical, train you in how to do it yourself. A guide from the Institute for Occupational Safety and Health (IOSH) lists the key risk factors to watch out for regarding the workstation, lighting, heating, ventilation and other issues (see Table 18.1).

The risks of developing a work-related injury to hands, wrists, arms, neck and back through using ill-adjusted equipment at home are high. The correct way to adjust your workstation is

Table 18.1 Telework risk assessment checklist

Risk Factor	Notes	Suitable: Yes/No
Location of residence		
Security of worker		
Security of visiting staff		
Entry and exit points		
Workstation		
Supplied equipment OK?		
Suitable size and adjustable?		
Set up correctly?		
Special requirements?		
Storage		
Room		
Size (11 cubic meters minimum recommended)		
Lockable?		
Weight loading of floor?		
Lighting		
Natural		
Room		
Glare problems?		
Heating		
Type / adjustability		
Electrical installations		
13-amp single-phase 240-volt AC supply?		
Sufficient sockets?		
Test wiring		
Training cables?		
Protection, eg circuit breaks?		

(Source: adapted from IOSH guide)

examined on page 87. If you believe you may be developing the symptoms of RSI (see page 61) you should speak to your employer or GP at an early stage.

Also bear in mind that power sockets need to be sufficient in number, to avoid overloads, with no trailing cables as tripping hazards. Make sure your employer has checked your power supply.

Laptops

A UNISON survey of 500 careers advisers using laptops found that two-thirds reported eyestrain, and more than half complained of headaches, and back or neck pain. The HSE suggests the use of a docking station to enable the use of a normal-size screen/keyboard. Or raise the laptop on blocks and use a conventional keyboard. Take precautions when travelling with a laptop against theft and possible assault.

Breaks

While for homeworkers it is sometimes difficult to resist the temptation of hanging out the washing, or daytime TV, the serious point is that breaks or changes of activity are essential to safe working at home. It is also important to ensure you separate home from work commitments, eg childcare, hours of work. Providing a separate phone line for work can help – switched to an answerphone at the end of the working day.

Personal safety

The Suzy Lamplugh Trust offers this personal safety and other advice:

- **Assess the risks**. Ask yourself, 'Would I be happy to let my best friend do this?' Make adequate arrangements to ensure you are safe at all times, especially when clients visit you, or when you go to meet them.
- **Find a 'buddy'**. The main safety issue for people working from home is to ensure that they let other people know whom they are meeting. Make an arrangement with someone you trust, whom you can fax or e-mail every day with an itinerary of whom you are meeting, where and at what times. Your buddy could, for example, be another person working from home.
- **Back-up**. Have a plan and practise it until you and your back-up partner are sure what is expected of you.

- **Emergency routines**. Have a way of calling for assistance – code words, panic buttons. Think what will work for you.
- **PLAN**. Make safe arrangements to meet clients, using the PLAN rules:
  - P: Plan to meet first-time visitors in a busy public place, rather than your home, if possible.
  - L: Log in your visitors with a buddy and phone after to let someone know you are safe.
  - A: Avoid situations that could be difficult.
  - N: Never assume it won't happen to you.
- **Do not ignore your instincts**. Have a few polite phrases prepared to turn away any client you feel 'wrong' about. If the client looks respectable but you feel uneasy, say you've double-booked and another client is already there.
- **Someone you 'know'**. It's worth remembering that many people are attacked by someone they know or think they know. It's all too easy to start thinking of a 'client' as a friend. It's safer to keep a professional distance and not allow clients to wander around your home. If they want to use the loo, keep a discreet eye on them.
- **When clients visit you**. Make a phone call after the visitor has arrived, telling someone at the other end of the line that you will get back to him or her at a certain time, after the visitor has left. This acts as both an information call and a deterrent.
- **What message is your home giving?** Are you obviously alone or are you giving a sense that others are around? Are you sending out businesslike signals and being clear about your boundaries?
- **Exits**. Make sure you can get out easily. Don't be trapped; keep yourself between your client and your exit route.

Problems of isolation

Working alone can involve a strong sense of isolation, a lack of support and increased working hours. Your employer should be providing regular contact, eg through newsletters, access to training, regular work reviews. The Telework Association, the main national organization for teleworkers, believes that contact between colleagues should be encouraged both within the teleworking team and also between teleworkers and workers in the main office, both on a work and social level.

The Telephone Helplines Association, representing the major voluntary sector helplines, highlights the following disadvantages of working at home alone:

- **Angry or abusive calls**. When faced with an angry or abusive call, workers in an office can be more readily supported and supervised. People at home may not be able to access the same sort of back-up. They may feel pressured to take the next call, rather than taking time out to deal with the emotions the previous call presented.
- **Sharing experience**. Working alone limits the opportunity to share experience and learn from others, about call handling techniques or about the services provided.
- **Socializing.** Work has a social aspect. Working alone limits the opportunity to develop new relationships, and to discuss work experiences off-site.

Further information

Contact The Telework Association (TCA) (tel: 0800 616008; Web site: www.tca.org.uk). Services include the regular monthly newsletter, *Teleworker*, and access to local groups.

Guidelines for Good Practice, Telephone Helplines Association (tel: 020 7248 3388), price £10.

Homeworking hazards, go to: www.healthworks-in-London.org.uk.

The Teleworking Handbook, available from the TCA (tel: 0800 616008), price £19.95.

Teleworking: Out of 'site', out of mind?, Budworth, Neil. Technical information sheet, IOSH (tel: 0116 257 3100; Web site (to download sheet): www.iosh.co.uk).

Working Alone in Safety and *Homeworking: Guidance for employers and employees on health and safety*, available from HSE Books (tel: 01787 881165; Web site: www.hse.gov.uk/hsehome.htm), free.

19 Men's health – the working years

Gender health gap

Men don't like going to doctors when they are ill. They don't examine themselves for lumps and bumps, and they are less likely to report the symptoms of disease or illness. Their condition is likely to have worsened by the time they decide to seek diagnosis. Men's life expectancy is at least five years shorter than women's. Official statistics show that men suffer the great majority of fatal or major non-fatal injuries at work (see Table 19.1).

In the past few years, organizations such as the Royal Society of Medicine (RSM) and the Men's Health Forum have begun to tackle health inequalities for men. That there is a massive 'gender health gap' between men and women was made plain to an audience of GPs and occupational health specialists at an RSM conference by Professor John Ashton. He is an adviser to the Government's 'Our healthier nation' strategy, and regional director for public health in the north-west of England. Professor Ashton highlighted:

- **Shorter life expectancy**. Women in the highest social class have a life expectancy of 79.5 years, but it is eight years less for working class men. And there is a 15-year gap in life expectancy between women in the highest socio-economic

Table 19.1 Fatalities and major injuries at work, 2001/02

	Men	Women	Total	Men, %
Fatalities	200	4	204	98%
Major injuries	20,607	6,870	27,477	75%

(Source: HSE)

groups and working-class men from the most deprived ethnic minority backgrounds.

- **Higher occupational risks**. Deaths from occupational causes are far higher for men than for women, as Table 19.1 shows.
- **Premature deaths and suicides**. As a result of a complex of pressures, death from accidents and suicides among young men are now four times higher than for young women.
- **Coronary heart disease**. Death rates are up to five times higher for men.

Some diseases, such as prostate and testicular cancer, are biologically defined as male problems, and will be dealt with as such. Prostate cancer is the most common cancer affecting men alone. Nearly 22,000 men in the UK are newly diagnosed with prostate cancer each year and about 9,500 die. The number of new cases diagnosed is expected to treble over the next 20 years.

But the underlying cause of this broader pattern of gender health inequality lies in 'differential exposure to risk factors', and is therefore preventable. These risks include 'lifestyle' issues such as diet, a reluctance to talk about personal matters and working in occupations with a poor safety record.

Dr Ian Banks, coordinator of the RSM's Men's Health Forum, says that employers and health service providers are not doing enough to tackle the built-in, macho mentality where men won't talk to one another, and won't go to their GP. He asked one male patient why he was there. The answer came: 'I don't know, the wife sent me.' It emerged that the man was suffering from depression after being made redundant.

Men's health initiatives

In September 2001, the Men's Health Forum launched the *Men's Health Journal*, a quarterly magazine aimed at health professionals interested in men's health issues. Its Web site includes an *A to Z* of male health problems, from asthma and back pain to prostate problems, stress and weight loss, and advice on tackling some work-related health conditions.

Some employers, and most local community health projects, are now addressing men's health inequality. These are some examples:

Tackling obesity

Almost one-fifth of adult males are obese. As a chronic condition, obesity requires lifelong treatment, similar to diabetes or hypertension. Occupational health specialists are now including waistline measurements in employee health assessments. Waist is as important a measure as weight. One health specialist suggests that 'anyone with a waistline exceeding 102 centimetres should be considering lifestyle changes', such as diet and physical exercise.

Workplace health programmes

A chemicals manufacturer achieved a high take-up of workplace health initiatives among their mainly male workforce, through well-publicised awareness-raising initiatives. The wellness scheme is just one part of the health and safety programme at the plant. It started with a voluntary epidemiological study of the workforce, which revealed that cancer and ischaemic heart disease (inadequate blood supply to the heart) were the two principal health risks. Accordingly, the wellness programme was adapted: its cornerstone is now a twice-yearly personal health assessment, undertaken by a fully qualified nurse:

- Tests last around two hours, during which employees complete a general health questionnaire covering stress, alcohol intake, diet and physical activity.
- Measurements include blood cholesterol, urinalysis and blood pressure. Further tests include flexibility and a computer-based fitness test.
- The session ends with a discussion on advisable lifestyle changes and individual responsibilities, with a GP referral if advisable.

Other elements in the wellness programme include health promotions, smoking cessation and subsidized access to local leisure facilities.

Jane Deville-Almond, an occupational nurse associated with the Men's Health Forum, carried out a free men's health-check weekend at Harley-Davidson, Wolverhampton. Most of the 52 men had a 15- to 20-minute full health MOT, including cholesterol, blood sugar and blood pressure readings, a body mass check, and discussion time for testicular, prostate and bowel awareness. She found that:

- two-thirds failed their MOT with one or more long-term health risk problems;
- one in eight did not know the name of their GP;
- around half were overweight;
- a quarter presented with high blood pressure.

Other problems raised included difficulty in getting GPs to take worries seriously, in particular where mental health was concerned. Several men discussed feeling low and depressed but felt there was no one for them to turn to. One man had presented to his GP with the symptoms of bowel cancer and his GP had advised him to eat more vegetables and stop worrying (this man had not visited his GP for three years prior to this visit). Since he had lost weight over the previous month and was suffering constant bouts of diarrhoea, with blood present, he was advised to go back and see his GP.

Men see their GP one-third less often than women, and are likely to turn up to a surgery much later in the course of any given illness. Taking health checks to the workplace is one way to raise male health awareness.

Further information

Men's Health Journal, www.menshealthforum.org.uk, and its associated Web site on male health: www.malehealth.org.uk.

Part 5

Resources

20 Disability rights at work

Compared with many EU nations, Britain has a 'lamentable record' on getting people back to work after a serious injury or illness caused, or made worse, by their employment. According to the TUC, employees with work-related ill health currently have a less than 1 in 10 chance of returning to work. The TUC wants the Government to introduce new legislation requiring employers to rehabilitate their workers who return after a period of work-related sickness absence. It believes this will help concentrate employers' minds on operating safer working practices. Gita's experience (Chapter 4) strongly supports the TUC's case.

In the UK now:

- 25,000 people quit the labour market each year because of work-related injury or illness.
- Each week, about 3,000 workers on long-term sickness absence move from Statutory Sick Pay (SSP) to Incapacity Benefit. Of these, 90 per cent never return to work.
- The average time off for work-related ill health is 36 days.
- Long-term sickness absences, defined as being off sick for 20 days or more, take up the lion's share (83 per cent) of the 18 million working days lost each year through work-related illness or injury.
- The cost of workplace adjustment for employers can be quite low: as little as £200 for workstation assessment; and nil for employees referred through their GP for NHS-based counselling or psychological support to manage anxiety levels or stress.
- There are over 6.5 million people in the UK with a long-term disability or health problem that has a substantial adverse

effect on their day-to-day activities or that limits the work they can do.

This chapter covers:

- the Disability Discrimination Act;
- who is covered by the DDA;
- 'reasonable adjustments' at work;
- Access to Work;
- disability leave;
- what you can do for yourself;
- taking legal action.

The Disability Discrimination Act

The Disability Discrimination Act 1995 (DDA) heralded a brand new approach to tackling disability discrimination. The DDA replaced the Disabled Persons (Employment) Act, which used to require organizations to employ a 3 per cent quota of disabled people. The law has since been significantly strengthened by the DDA 1995 (Amendment) Regulations 2003, which come into force in October 2004.

The new approach includes an *individual right* to non-discrimination on the basis of disability, so that only in very limited situations can an employer justify treating a disabled person less favourably, and a new *duty on employers* to make 'reasonable adjustments' to the physical features on premises, and arrangements for employing disabled people.

A study by Industrial Relations Services, *Fit for the Job: Health, safety and disability at work*, shows that the DDA has had a significant impact on managers and workers alike. The study revealed that, while organizations may need specialist advice, most adjustments:

- are common sense;
- are low-cost (up to £200) and require little imagination;

- require direct consultation with the disabled worker; and
- require, if relevant, consultation with the union.

Who is covered by the DDA?

The DDA covers all employers *except* those with fewer than 15 employees. But, from October 2004, the DDA will apply to all employers regardless of workforce size, including groups so far excluded, such as the police, firefighters, barristers, partners and people on work experience. The armed forces remain excluded.

The DDA prohibits unlawful discrimination against a 'disabled person' in employment. Section 1 of the Act says: 'a person has a disability for the purposes of the Act if he or she has a physical or mental impairment which has a substantial and long-term adverse effect on his/her ability to carry out normal day-to-day activities'.

Discrimination law specifically includes harassment and victimization. In Chapter 14 we include the new legal definition of harassment common to discrimination 'on grounds of race or ethnic or national origins, sexual orientation, or religion or belief, or for a reason that relates to a disabled person's disability' (see page 174). This takes effect in October 2004.

Taking each of these issues in turn:

- **Long-term**. This means that the disability must have lasted, or can be expected to last, at least 12 months or, indeed, for the rest of the person's life. People with a fluctuating condition that is likely to recur within 12 months are covered. People with progressive conditions, such as cancer or multiple sclerosis, are included. However, cancer is only included once it has had an effect, not automatically from its diagnosis alone.
- **Impairment**. Physical impairments are not listed in the Code of Practice. Recent court cases have applied the DDA to back disorders, ME (chronic fatigue syndrome), depression and asthma. 'Mental impairment' also includes learning difficulties. Dyslexia is also covered.

- **Normal day-to-day activities**. The Act lists the broad types of 'normal day-to-day activities' that it covers:
  - mobility;
  - manual dexterity;
  - continence;
  - ability to lift, move or carry everyday objects;
  - speech;
  - hearing;
  - sight;
  - memory;
  - the ability to learn, understand or concentrate;
  - the perception of risk or physical danger.

The list does *not* specify work tasks.

However, you should be aware that medical diagnosis does not automatically mean the Act applies to your case. In many Employment Tribunal (ET) cases, the employer's first line of defence is to deny that the claimant is 'disabled' in the terms of the DDA. So proper medical advice is essential unless the disability is not contested. The ET will treat each case on its own merits.

'Reasonable adjustments'

The DDA covers recruitment selection procedures, contractual arrangements and working conditions. The Government's Code of Practice on the employment provisions of the Act gives examples of 'reasonable adjustments' that employers will typically have to make. The IRS survey revealed that the most common adjustments made by employers were, in rank order:

- allowing the disabled person to be absent during working hours for rehabilitation, assessment or medical treatment;
- acquiring or modifying equipment;
- altering the disabled person's working hours;
- transferring the disabled person to fill an existing vacancy;
- adjustments to premises;

- allocating some of the disabled person's duties to another person;
- assigning the disabled person to a different place of work;
- giving the disabled person training;
- providing a reader or interpreter;
- providing support workers; and
- modifying instructions and reference manuals.

For instance, allowing a disabled person some flexibility over their working hours enables the person to, for example, take additional breaks to overcome fatigue arising from the disability, or change their hours to fit with the availability of a carer. Providing a different place of work could mean transferring a wheelchair user's workstation from an inaccessible third-floor office to an accessible one on the ground floor. It could mean moving the person to other premises of the employer if the first building is inaccessible.

Good practice guides

Excellent good practice guides are available from voluntary organizations covering differing impairments, the RNIB (which publishes the *Get Back!* booklets), trade unions and the Employers' Forum on Disability. These guides cover general adjustments, as well as offering specific advice for employees with a wide range of impairments, for example:

- mental health;
- diabetes;
- upper limb disorders and RSI;
- visual impairments;
- hearing impairments;
- progressive conditions;
- dyslexia.

The DDA Code of Practice on employment provides examples of adjustments that should be considered by an employer for blind and visually impaired people.

The RNIB *Employment Factsheet* highlights the following advice:

- providing practical support to visually impaired people in the form of personal reader support, or adapted computers with large character, Braille display or speech output;
- reallocating some minor duties to another non-disabled employee;
- providing practical support when an employee becomes visually impaired, so that he or she has a fair chance of retaining employment after assessment and rehabilitation.

Access to Work

The Access to Work (ATW) scheme, run by the Employment Service, provides financial assistance towards the extra costs of employing someone with a disability. It is available to employed, unemployed and self-employed people, and can apply to any job, full- or part-time, permanent or temporary.

Support available includes:

- a communicator at job interviews for people with a hearing impairment;
- a reader for someone who is blind, or has a visual impairment;
- special equipment, or alterations to existing equipment;
- alterations to premises or to the working environment; and
- travel-to-work costs.

Applications for support are available from the Disability Employment Adviser in the Disability Support Team at your local Jobcentre. If you are unhappy with the support provided, you can ask the support team to review its decision. Outside agencies, such as the RNIB, provide advice on adaptations, and on appeals. Key provisions of Access to Work are summarized at the

Employers' Forum on Disability Web site: www.employers-forum.co.uk/www/guest/info/factsheets/sheet1.htm.

The *GetBack! Pack* is a series of disability-specific guides published by the RNIB, but useful for any disability. It includes the *Employee's Guide and Checklist* for people who, due to accident, ill health or disability, are finding it difficult to do their job, but would like to remain in work. The guide is available free on the RNIB Web site. The checklist is 'to help you think through your circumstances, and prepare to discuss them with your employer, and, if necessary, with an external employment adviser'.

The guide helps you to:

- clarify your expectations;
- identify what work-related activities you can and can't do;
- analyse your job: difficulties and solutions;
- get help if needed;
- prepare to meet your employer.

Disability leave

Some unions have negotiated special 'disability leave' for employees, defined by the TGWU as 'time to adjust to their changed circumstances, including paid leave'.

The aim of disability leave is to give time and job security to disabled employees. The leave allows them to assess whether they would be able to resume work following adaptations to the premises or to their workstation, by learning new skills or by doing their job in different ways. The TUC supports a statutory right for employees to have up to one year's paid disability leave, and encourages unions to negotiate paid time off work with employers. Examples of union models for agreements include the TGWU's model disability agreement and the advice of the Council of Civil Service Unions (CCSU).

The TGWU's model disability agreement includes counselling and support services, and other opportunities for employees to discuss their situation with their line manager, personnel manger,

OH staff, union representative and 'other specialists nominated by the union'.

Similarly, the civil and public services union's guide is based on a two-year study of disability leave by the RNIB. The free booklet, *Disabilities: A negotiator's guide*, emphasizes that 'any assessments must include the views of the person with the disability – and provide the opportunity for them to be represented by their union'. Disability policies should be written and their details made known to staff. All the options should be considered, in the following sequence: adapting the employee's existing job to suit his or her new situation; alternative posts combining current and new duties, which an employee can deal with effectively; or the offer of a transfer to new work, with pay protection where necessary.

What you can do for yourself

If you are suffering from an injury caused, or made worse by, your job, and you want to return to work, the essential questions are:

- Can you return to your old job, without any changes?
- Can you return to your old job with minor changes, perhaps for a short time, eg to your working hours, some of your duties, etc?
- Can you do your old job, with significant changes, to your workload, where and how you work?
- Do you need to transfer to suitable alternative employment?
- **Your employer**. Your employer is legally obliged to make 'reasonable adjustments' to enable you to return to work (see page 222). What may be considered 'reasonable adjustments' will depend on the circumstances of your case. Key points:
  - Obtain your organization's *written procedures* (if any) on supporting people with a disability.
  - You must be *fully consulted* and involved in the adjustments process at every stage.
  - Many employers prepare a *written agreement* with their employees as the basis of a return to work.

- **Early intervention** can greatly increase the likelihood of a successful return to work. The longer you are left to languish at home, the less likely it is you will return to your former job at all. This is especially the case, for instance, with back injuries and other musculo-skeletal disorders. This is one reason why some GPs take quite a robust attitude to their patients who have been off sick for several weeks. They see the dangers in not getting back in contact with your employer after weeks of absence.

- **Your GP**. You will need to talk through these options with your GP. The greater the changes needed, the more contact there will be with your employer, through your GP. Some GPs know more about employers' duties to make 'reasonable adjustments' and about good practice than others. Work 'adaptations' is a very broad term, ranging from a few basic changes in your start time, days of work or responsibilities, to significant changes, eg providing disabled access. You should also read through the GP case studies in Chapter 21.

- **Your union rep**. A union rep should have access to expert help and advice on health, safety and legal issues. Your union may have knowledge and experience of handling similar cases with your employer, or elsewhere, plus experience in representing people at work, and independence from the employer. You have the right to take someone with you when you meet your employer (see page 271).

- **Voluntary organizations**. There is a national voluntary organization for almost all of the main types of work-related injury or ill health. They offer advice, support and information. Obtain a copy of *Get Back! For Employees*. (For contacts, see Further information, below.)

- **Your employer's OH service**. If your employer has an occupational health service, it will be able to advise you on your return-to-work options, 'readjustments' and other issues (see Chapter 22). Key points:
  - The OH service should not release confidential information on your health condition to your supervisor/line manager without your consent.

- The OH nurse or doctor may want to contact your GP. This can only happen with your consent. Your GP will want to discuss his or her reply, and can only respond with your agreement.

Taking legal action

By its very nature, long-term absence is very likely to stem from a disability. Many of the over 9,000 disability claims lodged with Employment Tribunals since the DDA was passed have involved workers dismissed from their job after long-term absence.

Legal cases show that the test of *disability discrimination* can be tougher on an employer than *unfair dismissal* itself. One important tribunal case involved a health worker dismissed after a long period of sickness absence. On the one hand, the tribunal ruled against the worker, Mrs Ridley, for her unfair dismissal claim, arguing that her employer had taken reasonable steps to find her alternative work. But they *upheld her disability discrimination claim* on the basis that the DDA imposed tougher obligations on employers to make reasonable adjustments, or find other work, for a person with disabilities. Mrs Ridley's employer had sent her a list of vacancies, but had made no other attempt to find her alternative work (*Ridley* v *Severn NHS Trust*).

To make a tribunal claim, you will need specialist legal advice, available through your union, local law centre, citizens advice bureau or other source (see Chapter 24). In 2003, the average tribunal award was £9,800 for financial losses and £3,600 for injury to feelings.

'New vision' for occupational health

The Government wants more employers to rehabilitate workers suffering from injury or ill health. This includes people who have sustained their injury outside the workplace, eg through sport or traffic accident.

In practical terms, the new initiative includes:

- improving occupational health services in GP surgeries (see page 252);
- NHS-Plus: a scheme to encourage NHS hospitals to provide occupational health services to smaller firms with no in-house OH nurse or doctor;
- job-retention pilot schemes across the UK, aiming to provide employment, benefit and health support when people have been on Statutory Sick Pay for a period of six weeks.

Further information

Disabilities: A negotiator's guide, available from the Public and Commercial Services Union (tel: 020 7801 2683), free.

Employers' Forum on Disability (tel: 020 7403 3020; Web site: www.employers-forum.co.uk).

Get Back! Pack, including *Employees Guide and Checklist*, available from RNIB/Rehab UK (tel: 0845 702 3153), price £2.50. Details of other guides on request.

How to Survive Working Life, available from Mind (mental health charity) (tel: 08457 660163; Web site: www.mind.org.uk).

Repetitive Strain Injury Association, a national support group for people with upper limb disorders (Web site: www.rsi-uk.org.uk).

Royal National Institute for the Blind (RNIB) factsheets cover the DDA, the Access to Work scheme, retaining your job and other issues. The booklet, *Adapting to Change*, is available from tel: 020 7388 1266; Web site: www.rnib.org.uk.

The Disability Rights Commission (DRC) provides advice and information on all aspects of the employment of disabled people. DRC helpline (tel: 08457 622633; e-mail: ddahelp@sta.sitel.co.uk; Web site: www.drc-gb.org).

The Government's official disability Web site has free copies of the Disability Discrimination Act, the Code of Practice for the elimination of discrimination in employment against disabled persons, and the DDA Regulations taking effect in October 2004 (Web site: www.disability.gov.uk).

21 What to expect from your doctor

About 1 in every 12 patients visiting a GP is seeking treatment or advice on a work-related health problem. No matter where you work, the sympathy and support of your GP may be the only independent, professional medical help you can rely on. Most of us are not covered by a workplace occupational health service.

But GPs' knowledge and experience of occupational healthcare varies widely. Some are much more alert to occupational health issues than others. For the typical GP with 2,400 patients on the list, the time taken up liaising with an employer over a single patient's work-related illness has to compete with many other priorities. Since December 2000, a Government initiative has been providing occupational health training for GPs, aimed at improving their understanding of the effects of work on employee health.

This chapter aims to help you to:

- understand a GP's approach to any patient suffering from work-related illness;
- put your points across to your GP;
- provide information that your GP might overlook;
- build self-confidence in dealing with your GP;
- ask, 'What do you mean?' if your GP uses terms you do not understand.

We include:

- questions your doctor should ask about your work;
- workplace hazards: what GPs should know;

- seven GP case studies: pregnancy and health; 'tennis elbow'; work-related upper limb disorders; lower back pain; occupational asthma; latex allergy; and workplace bullying.

Questions your doctor should ask about your work

'The first question I now ask when I see a new patient is, "What is your job? What do you do? Where do you work?" For me, it's now important'. This GP is one of many in Leeds working closely with the Leeds Occupational Health Project – one of a number of community-based occupational health services in the UK (see Chapter 23). If doctors or nurses are not to miss important occupational factors in their understanding of your ill health or injury, they need to ask you about your work, past and present.

Questions your GP should ask include:

- What is your job?
- What do you actually do at work?
- Do you have any other jobs or hobbies that may be relevant?
- What previous jobs have you done?
- What hazards are you exposed to now?
- What are your hours of work/shifts?
- Do your symptoms change (get better or worse) during the working day, at weekends or on holiday?
- What kinds of pressures do you work under?
- What do you think is the cause?
- Does anyone else at work have the same problem?

Be sure to tell your GP about the tasks you actually perform at work. Describe the pressures you work under and the hazards you are exposed to. This is essential to help your GP assess the real causes of your condition.

In some cases, it will be necessary to go back over your work history to identify a work-related cause of ill health. Noise-related deafness may have started in an earlier job, for example. Be prepared to *offer this information* to your GP. If your doctor does not ask these kinds of questions, and you believe them to be relevant, you need to say so at the time.

Workplace hazards: what GPs need to know

Your GP will need to understand the types of hazard you are exposed to at work, simply because ill health or injury arise because of exposure to hazards. These could be:

- physical;
- mental;
- chemical; and
- biological.

The words 'hazard' and 'risk' do not have the same meaning. A hazard is something capable of causing you harm: a substance, work practice or situation. Risk is the likelihood of the harm actually occurring.

'Risk assessment' of a hazard is the fundamental basis of health and safety practice in the workplace, and is the basis of most modern legislation.

Health-at-work surveys

Employers are required to survey the health effects of these kinds of hazards. If unusual results are found, the employer's occupational physician (if there is one) will ask the worker for permission to inform the worker's GP of the results, so that the GP is fully informed and able to follow up with any necessary medical treatment. Equally, with your consent, your GP can write to ask for the information about any survey in which you have taken part, or which covers the type of work you do, regardless of how the employer has used the results.

Advice or treatment?

How your GP proceeds after the questioning, the health check and diagnosis depends on the nature of the illness, the treatment you need and the likelihood of an early return to work in the same job.

GPs have a vital advocacy role to play in helping patients deal with work-related injury or ill health. Many patients do not have union representation or a safety rep at work. A well-informed GP can provide vital advice and support to employees and, with the patient's consent, to their employer.

In *Occupational Health Matters in General Practice,* a practical guide for GPs, Dr Ruth Chambers and colleagues point out that doctors cannot contact an employer without their patient's consent. Nor are they obliged to give advice to employers in relation to occupational health issues.

A GP has a number of issues to consider and share with the patient:

- **Is further diagnosis needed?** including hearing tests (audiometry), lung function tests (spirometry) and other more specialist examinations.
- **Treatment**: including medication, possible referral for physiotherapy or counselling, referral for X-rays or to a specialist.
- **A sick note**, or perhaps several, to enable recovery away from work.
- **Return to work: what are the options?** In support of your return to work, your GP may, with your consent, write to your employer describing your condition and suggesting the steps your employer should take. This may be things your employer must do to comply with health and safety Regulations, or suggested steps to help facilitate your return to work.
- **Confidentiality**. Any report provided by a GP to the employer needs to comply with the requirements of the Access to Medical Records Act 1988, which gives the patient rights of consent/refusal, access and comment.

- **Wider patient knowledge**. Subject to your consent, a GP can provide an employer with an assessment of the wider family and social issues involved in a case of work-related ill health, from his or her knowledge of the patient's circumstances.

Seven GP case studies

The following health scenarios show how a GP might tackle specific health problems. They are adapted from our interviews, from cases in *Occupational Health Matters in General Practice* and other sources.

Case study 1: Pregnancy and work-related risks

Eileen, who is 16 weeks pregnant, attends her GP surgery. She works in a local chemical factory, and consults her GP because she is worried about the effects on herself and her unborn child of exposure to chemicals.

The GP might consider these matters:

- Find out precisely what chemicals the woman works with, or might be exposed to. Even if the woman is able to provide the list, the names and their effects may not mean much to the GP. The GP may ask the HSE or local authority environmental health service for advice on the health risks associated with these chemicals. Some chemicals may cause cancer; others may cause genetic damage or risk to the unborn child or breastfed baby.
- If the employer has an OH service, it will have the information on the chemicals used, and perhaps on the specific risks to a pregnant woman.
- The law requires an employer to assess the risks of chemicals to all employees who are pregnant or who have given birth within the previous six months, or who are breastfeeding, and to do what is 'reasonably practical' to reduce the risks. See the HSE leaflet, *A Guide for New and Expectant Mothers who Work* (www.hse.gov.uk/mothers).

- It is the employer's responsibility to decide on whether the risks to the mother or unborn child warrant changes to the woman's work routines, eg redeployment to an alternative job or paid leave for as long as is necessary.
- If the employer has an OH service, the GP can check to make sure that these steps are being taken.

Workers with certain medical conditions, such as epilepsy, diabetes, pregnancy and, in some cases, PMS (see Chapter 16), need special care in terms of fitness to work.

The Management of Health and Safety at Work Regulations 1999 require employers to take steps to protect completely any pregnant woman against exposure to chemical or biological agent, work process or working situation that may cause harm to the woman or her unborn child (see page 35).

Case study 2: 'Tennis elbow'

'Tennis elbow' (medical term: 'lateral epicondylitis') is common among workers who put strain on their forearm muscles by using hand tools, or who work with their forearm flat ('pronate') and the elbow partly bent. Sufferers include plasterers, joiners and brick layers, and computer operators, because of the need to grip the mouse with the forearm pronated and to move the mouse around. This puts strain on the insertion of the tendons into the lateral epicondyle.

The GP should ask when the symptoms started and whether the patient thinks they are made worse by work. The GP is also likely to ask about hobbies and sports (including tennis!). Depending on the severity of the condition, the GP will then discuss the options with the patient, including: treatment, perhaps with a referral to a specialist or to physiotherapy; and alternative work. To aid recovery, it may be advisable to avoid any of the tasks that contributed to the condition. If no alternative work is available, then a period off work may be needed. However, if the employer has an OH service, the GP may, with the patient's agreement, contact the OH service, outlining the condition and suggesting a period of alternative duties.

Case study 3: Work-related upper limb disorders

Claire works as a data entry operator for a large local company. She visits her GP to complain of aching and shooting pains in the back of her hands, with her right hand worse than her left. Sometimes, the pain radiates to the right forearm and shoulder. She has suffered sleepless nights. The pain used to ease off over the weekend, but it is now more or less continuous. She says that the pace of work has increased in recent months.

The GP asks about the type of work she does, her workstation layout, the equipment she uses and how her daily workload has changed recently. It emerges that the symptoms come and go, but she is definitely getting worse. The doctor's examination reveals tenderness of the right common extensor muscles.

Claire asks the GP to send a letter to her management, requesting a change of work to help ease her symptoms. The GP agrees and his letter briefly describes her condition, informs the employer about the referral to a specialist and asks whether it would be possible for the employer to change Claire's work in the meanwhile so that she is not continuously working at a keyboard, to help keep her symptoms in check. He copies the letter to Claire, and refers her to a rheumatologist.

But the company replies that it is in 'full compliance' with the Display Screen Equipment (DSE) Regulations. Workstations are checked periodically, including Claire's. They enclose a staff handbook which says that staff take regular breaks away from computer screen work, by doing other tasks, such as filing. The letter also adds that Claire has taken 10 single days off sick in the past year, and none of her self-certificated notes refers to pain in the hands or arms. Her attendance record has, therefore, caused concern.

The rheumatologist writes to the GP stating that: 'Claire clearly has a repetitive strain injury, and should be absent from work for some time, since the condition is so aggravated by it'. He has arranged for physiotherapy and, with Claire's permission, has already written to the company.

At her next GP appointment, Claire says that she had taken occasional days off work with wrist pain, which used to clear up quite quickly. She was reluctant to let her employer know the true reason at first. She says there was usually little alternative work to do away from her workstation. Now, she wants the problem resolved, with a course of physiotherapy and a transfer to alternative work. With Claire's consent, the GP writes a further letter to her employer requesting a transfer. The GP has carefully supported Claire's dealings with her employer. The employer is now faced with a straight choice.

Case study 4: Low back pain

Gwen, a mother of four, worked as a packer in a food factory. She attended surgery complaining of a bad back and, on examination, appeared to have a problem. Her symptoms included muscular tenderness adjacent to the lower vertebrae, and limited straight-leg-raising ability on the right-hand side. She was overweight. She asked the GP to write to her employer to transfer her to lighter duties.

The GP wrote to Gwen's employer, as she requested. The employer wrote back agreeing to transfer her to the label control department, where there was no heavy lifting. Some months later, Gwen returned to the surgery. She wanted to return to her former job, because her back had recovered, she felt that her new job had set her apart from her fellow workers and she had lost the opportunity of overtime work. Her earnings had fallen considerably.

The GP wrote to her employer, but this time the company replied pointing out that she had well understood the move to be permanent. In any case, even if she missed the overtime in her former job, the company would not necessarily have agreed to her working extra hours, as a precautionary measure. Gwen's case had reached stalemate.

Managing lower back pain

Over two-thirds of adults experience low back pain at some time. GPs are familiar with the condition, but their standards of diagnosis and treatment vary widely.

Dr Ruth Chambers believes that: 'For any worker who develops acute low back pain, the ways in which the GP manages both clinical care and occupational issues are extremely important'. If the GP is satisfied that there is no underlying spinal problem, then dealing with the *psychosocial* issues and the *attitude towards returning to work* becomes as important as managing the pain and immobility.

Key issues for the GP to consider include:

- If the patient works in a heavy, physical job, it may be necessary to explore alternative, lighter duties for a while, to help him or her return to work. 'This should be accompanied by active encouragement to deal with the psychological barriers to returning to work, and help with learning to manage pain,' Dr Chambers suggests.
- For workers in lighter physical jobs, early return to normal work is now often preferred by GPs. But care needs to be taken in pain management.
- The longer the period off work, the higher the chances that an employee will not return at all. The crucial period is between 4 and 12 weeks, after which the chances of being off for a full year are quite high.
- Dealing with fears, such as that 'pain is harmful', or with the low mood associated with prolonged absence and some pain, is as important as dealing with the pain itself.

But a typical patient with work-related back pain might wait six to nine weeks for specialist treatment, by which time his or her chances of returning to work will have halved. According to Dr Harry Waldron, an occupational medicine consultant, the best chance of recovery lies in having 'active rehabilitation' and going back to your job as soon as possible, with proper supervision by an occupational health specialist.

Understanding Back Pain, a booklet in the Family Doctor Publications series, provides a useful supplement to advice and treatment provided by a GP.

Case study 5: Occupational asthma

Chris, an asthma sufferer, visits his GP having inhaled the fumes resulting from a chemical spillage at work. He had been admitted overnight to hospital as a result, and brings with him the discharge letter from the hospital. This letter includes results of breathing tests ('spirometry') undertaken in the hospital. Chris says that his asthma has worsened since the incident.

The GP examines him and finds that his lungs sound clear. Chris himself says that he has no symptoms now. The GP should then ask two sets of questions: about the spillage incident, and about the asthma itself.

In response to the doctor's questions about his job, it emerges that Chris works as a paint sprayer using isocyanates, the chemicals found in two-pack spray paints. The GP then asks about the frequency of the asthma spells, and whether they occur at weekends or holidays (time away from work), or follow a shift pattern.

The GP ascertains the nature of the chemical concerned in the spillage. It is an ammonia concentrate, pungent in odour, kept in a tin with a loose-fitting lid, accidentally knocked over.

Factors pointing to occupational asthma include that it appears *after* the work believed to be inducing it. Evening chest tightness and recovery by the next morning is a common pattern. As the problem becomes more severe, the asthma may encroach on the working day. The absence or reduction of symptoms at weekends or on holiday is another pointer to an association with work. Various spirometric tests, in the doctor's practice or in a hospital, over a few weeks, will establish when the asthma kicks in and subsides.

These tests may point to a strong association with work. If recognized early enough, occupational asthma can be tackled effectively by simply avoiding exposure. But this is not always possible, especially in the case of a skilled worker, such as Chris, whose livelihood depends on it.

The HSE's Code of Practice on Asthma (2002) requires employers to take steps to prevent, or adequately control, exposure to all substances likely to cause asthma. So if complete *avoidance* is not possible, then *controls* need to be introduced to bring down the level of exposure. Options include:

- use of a suitable and effective mask or protective hood;
- treating the symptoms through medication; and

- ensuring that the chemical's 'occupational exposure limit' (see page 16) is complied with: this involves ensuring that the employee is only exposed each day or shift to the extent permitted for that chemical.

Medical treatment and providing suitable personal protective equipment are relatively simple to achieve. Controlling exposure to the hazard requires solving the problem at source, for example, by substituting safer paint sprays. But with many smaller employers this may not happen.

Chris may continue in the same employment despite the risks to health, holding on to his skilled and well-paid job, with the asthma symptoms controlled through medication. If he is compelled to leave the job, or transfers to a less well-paid occupation, he may be able to claim compensation (see Chapter 24).

Understanding Asthma, in the Family Doctor Publications series, provides a useful supplement to advice and treatment provided by a GP. The booklet includes an illustrated anatomy of the respiratory system, a step-by-step guide to the diagnosis a GP or hospital is likely to recommend and a checklist of the common causes of occupational asthma (see Table 21.1).

Table 21.1 Common causes of occupational asthma

Causes/Substance	Occupations
Isocyanates	Painters, varnishers and some plastics workers
Colophony: a resin used in chemical processes, varnishes, glues, soaps and soldering agents	Solderers
Enzymes	Detergent production, drug / food technology workers
Flour	Bakery, catering trade workers

(source: *Understanding Asthma*)

Case study 6: Latex allergy

The cause of latex allergy is rubber protein, found in cheaper rubber gloves made from poorly washed latex. Latex allergy can produce severe reactions in sensitized individuals. Diagnosis is largely clinical, and involves referral to a specialist.

The treatment option is simple: avoidance. This is not easy if the affected person is a laboratory worker or a nurse, or if the cause is not identified early on. 'Hypoallergenic' gloves made from well-washed latex are an alternative.

Diane, a former staff nurse at Cardiff Royal Infirmary, was forced to give up her job after developing severe contact dermatitis. In 1990, Diane developed widespread eczema on both hands. She says:

> After 18 months working at the hospital, the skin on both my hands practically sheared right off. I went back and forth to my GP, until I eventually had skin tests ['patch tests'], which found that I was allergic to latex and nickel. I told my manager at the hospital but they refused point-blank to provide me with an alternative type of glove.

She continued working, and was susceptible to picking up infections from patients. Her right hand swelled to twice its normal size.

A member of health services union UNISON, Diane contacted her shop steward and was put in touch with the union's legal services. A personal injury claim was launched in the courts. In December 2000, she received a £100,000 compensation settlement. Diane says: 'Eczema is a very depressing condition. I lost my career. I am relieved that the case is settled, but I would rather have my health and job back'.

Understanding Skin Problems, in the Family Doctor Publications series, suggests that about 1 in 10 consultations in general practice concern skin diseases. 'Eczema' and 'dermatitis' mean the same thing: inflammation of the skin. The booklet describes the two main types of work-related dermatitis.

Allergic contact dermatitis is acquired from contact with perfumes, cement, nickel, rubber chemicals and leather. Treatments include damping down reactions with a cream (eg corticosteroid creams). 'A far better way is first to identify the chemical in question by patch testing, and then to avoid contact with it.' The testing of small, unaffected areas of skin (usually on the back) may take place in the doctor's surgery or hospital.

Irritant contact dermatitis accounts for most industrial cases. Prolonged exposure, sometimes for many years, is needed for weak irritants to cause this type of 'wear and tear' eczema. Cleaners, car mechanics, hairdressers and nurses are especially at risk. Irritants include: detergents, cleaning materials, cutting oils, alkalis and rough work. 'Treatment must include changes in the daily routine to reduce skin damage. Protective gloves should be worn wherever possible.'

Your employer's duties to prevent exposure to hazardous substances are discussed in Chapter 1.

Case study 7: Bullied at work

For this handbook, we interviewed Cath Noonan, Maxine, Teresa and other workers who suffered from bullying by supervisors and fellow workers. Each sought help from their GP.

For many targets of bullying, especially if there is no one to confide in at work, their GP becomes the first and chief source of help. Most family doctors prefer to deal with stressful experiences, such as bullying, by offering counselling and general advice. Just giving a full account of the circumstances is usually beneficial. It helps to get things in perspective, so that you are better able to begin to find a way forward.

Your GP is likely to try to help you focus on the likely causes of your distress. For Teresa, the sources of stress lay at work, and needed resolving there. But for many people, domestic and personal relationships also play a part.

A GP is likely to discuss your: **physical symptoms**, such as muscle tension, change in appetite, weight loss, sleep problems, bowel disorders; and **emotional reactions**, such as feeling under pressure, feeling tense and unable to relax or sleep, irritability, increased tearfulness, depression.

The GP may advise you to take some time off work and issue a sick note. This buys time and peace of mind. The GP may consider medical treatment, eg an antidepressant or counselling.

If your GP practice is linked to a community-based OH project (see page 252), you may be referred to a specialist, who can advise and support you in coming to terms with the bullying you are experiencing. After a period off work, your GP is likely to present you with two choices: returning to work or leaving the job.

You may feel a lot better in yourself, even though the situation at work may not have changed. Many GPs recognize that going back to work is made easier if union support and advice is on hand, or a sympathetic co-worker or manager. If your GP has been contacted by your employer, perhaps by its OH service, your doctor will discuss the return-to-work options being suggested. Your GP is also likely to go over your choices in dealing with the bully: confront the bully, formally or informally (see page 168); speak to your union, or to personnel; or make a formal complaint.

Leaving the job might be the more difficult option, and your GP may well advise against it. It is more difficult to find another job if you don't have a reference, or have been off sick for a long period. You may also decide to take legal advice on harassment at work (see page 172). Ultimately, it is your decision.

Understanding Stress, in the *Family Doctor* series, helps you to understand the physical and emotional signs of stress, how to cope and where to find help and support (see Further information, below).

Jobs at risk

Table 21.2 shows jobs at risk and the possible causes of the associated conditions.

Table 21.2 Jobs at risk, health conditions and causes

Condition	Jobs at Risk	Possible Causes
Respiratory diseases		
Asthma	Laboratory work	Small animals
	Car paint spraying	Isocyanates
	Healthcare workers	Glutaraldehyde
	Electronics, soldering	Colophony (a resin used in chemical processes)
Rhinitis	Healthcare workers	Glutaraldehyde
Skin diseases		
Contact dermatitis (irritant)	Domestic work	Detergents
	Building trades	Cement
	Painting	Solvents, polymers
Contact dermatitis (allergic)	Wearing rubber gloves	Latex protein
	Cleaning	Chemicals, sprays, etc
	Hairdressing	Chemicals, etc
Cardiovascular disease		
Raynaud's phenomenon	Use of hand tools	Vibration
Infectious diseases		
Hepatitis B	Healthcare work	Blood/body fluids
Ears, eyes, throat		
Hearing loss	Heavy industry, road mending	Noise
	Call centre work	Sudden, sharp noise
Eyestrain	DSE use	
Voice loss	Teaching	Excessive, inappropriate use/stress
	Call centre work	Excessive, inappropriate use/stress
Musculo-skeletal problems		
Tenosynovitis	Keyboard use	Prolonged, repetitive movements
	Checkout work	Prolonged repetitive movements
Epicondylitis ('tennis elbow')	Computer work plastering, bricklaying using hand tools	Repetitive movements
Low back pain	Labouring, heavy or light manual work	Lifting and twisting
Psychological problems		
Stress	Most occupations	Work overload, job insecurity, lack of control
Anxiety, depression	Most occupations	Work overload, job insecurity, lack of control

(Adapted from *Occupational Health Matters in General Practice*)

Further information

A Guide for New and Expectant Mothers who Work, HSE, free leaflet available from tel: 08701 54550 and Web site: www.hse.gov.uk/mothers.

Chambers, Ruth *et al* (2001) *Occupational Health Matters in General Practice*, Radcliffe Medical Press, 18 Marcham Road, Abingdon, Oxon OX14 1AA, price £18.95.

Booklets in the *Family Doctor* series:

- *Understanding Alcohol*, Chick, Dr Jonathan.
- *Understanding Asthma*, Ayres, Professor Jon.
- *Understanding Back Pain*, Jayson, Professor Malcolm.
- *Understanding Depression*, McKenzie, Dr Kwame.
- *Understanding Skin Problems*, Colver, Dr G and Savin, Dr J A.
- *Understanding Stress*, Wilkinson, Professor Greg.
- *Understanding Your Bowels*, Heaton, Dr Ken.

These booklets are available from pharmacists, supermarkets or direct from Family Doctor Publications (tel: 01202 668330), price £3.50 each.

22 Occupational health at work

Most public sector organizations, and many large private firms, provide a workplace occupational health (OH) service staffed by a doctor, a nurse and possibly other specialists, together with an on-site treatment room and other facilities. This chapter is about what to expect from a workplace OH service.

You are likely to be contacted by your employer's OH service if you are absent sick for more than two or three weeks, or if you expect to return to work without fully recovering from the effects of work-related ill health or injury.

Many large employers offer OH services. But, if you work for a small or medium-sized private firm, there's a less than 1 in 10 chance that your employer provides OH facilities.

None of this means you can't visit your GP. For medical and specialist occupational health advice, you are still best advised to visit your doctor (see Chapter 21), or a local community OH project, if there is one near you. One issue we look at in this chapter is the relationship between GPs and workplace OH services.

What is occupational health?

Occupational health (OH) is about the effects of work on health: whether through sudden injury or long-term exposure to hazards. It involves preventing work-related diseases, through safe working practices, ergonomics, health surveillance of the work-force and good management. Good OH practice is about adapting the work to the worker and not the other way round.

Employers' OH services help them to fulfil their statutory duties to ensure a safe and healthy working environment, undertake risk assessments, carry out health surveillance, protect the vulnerable and employ people with disabilities.

A fully fledged OH service will include an on-site clinic staffed by an OH doctor, an OH nurse and other health specialists. Some employers contract out their OH service to a private company or local hospital. Some buy in specialist support, eg physiotherapy.

When available, OH services can play a crucial role in your return to work after ill health or injury.

Specialist support provided by employers, directly or contracted out, includes physiotherapist, counsellor or occupational hygienist responsible for monitoring and maintaining environmental hazards such as air quality. Some organizations provide an Employee Assistance Programmes (EAP). These cover around 2 million workers in Britain, providing a telephone helpline, access to counselling and advice on personal and work-related issues. Some EAPs provide family cover.

A growing number of organizations also support healthcare cash plan schemes. These are provided by not-for-profit organizations such as Westfield and HSA. In return for small weekly contributions (starting at around £2 a week), providers meet all or part of the cost of everyday healthcare essentials, such as dental and optical treatment, physiotherapy and other specialist healthcare. They define their role as supplementing, not competing with, the NHS. Employers support these schemes because they encourage their staff to keep in good health. Individuals join them, with the greatest use made of the dental and optical benefits and refunds on expenses.

Ethics and confidentiality

OH doctors and nurses are bound by a duty of confidence towards their patients, enshrined in the ethics of their professions. The professional body for OH specialists is the Faculty of Occupational Medicine. Its *Guide to Standards in Occupational Medicine* requires members to 'keep all individual medical information confidential, releasing such information only with the individual's informed consent, or when required by law or overriding public interest'.

If your employer wants access to your occupational health records, then it must obtain your written agreement, or apply for a court order to gain access.

Specialists, such as occupational health psychologists and physiotherapists, are required to belong to their national professional association. These bodies regulate the conduct of their members through professional and ethical standards. These standards include client consent, confidentiality, anti-discriminatory practice and personal conduct.

Conflicts of interest are recognized in an *Occupational Health Advisory Committee Report* in 2000, which said: 'Even when occupational health support is provided by employers, it is often viewed with suspicion by workers. A recent survey of union reps by Labour Research Department found that many still see occupational health as a management tool, too concerned with sickness absence monitoring which could be used against staff'. The report concluded that a 'priority need' for workers is for occupational health support that is 'seen to be objective and independent of undue employer or management influence, is ethical, and of the highest probity'.

As Dr Ruth Chambers and colleagues point out in their guide, *Occupational Health Matters in General Practice*:

> Any doctor or nurse dealing with occupational health issues must combine the need to maintain strict medical confidentiality in relation to an individual employee, with the need to provide useful advice to the employer. This balance is still misunderstood by many employers, employees and doctors, who believe that a doctor or nurse may adopt a lower standard of medical ethics in relation to occupational health issues.

Confidentiality – points to watch

- **Consent**. All communications that relate to confidential medical information about 'a named employee' should be with the employee's 'informed and written consent'.
- **Medical details**: should not be given to managers without the employee's consent. Advice on an individual's fitness for work can usually be given 'in general terms' only.

- **Communications**. A GP or practice nurse communicating with a patient's employer should always try to make contact through the employer's OH department (nurse or physician), rather than releasing medical details direct to line or personnel managers.
- **GPs.** A GP has a vital advocacy role to play in relation to their patient's health at work. Many patients do not know how to use the health services effectively, and do not have a strong voice or representation at work.
- **OH doctor**. The occupational physician is not the employee's 'personal medical attendant', and has therefore to be careful to give advice to both the employer and the employee that is as fair and objective as possible.

Rehabilitation

Rehabilitation means the *process* of getting yourself back to work, either to the same job or to an agreed alternative that is appropriate to your medical condition. According to the TUC, there is a growing recognition that employers need to make more effort to retain workers who have been affected by poor health, injury or disability as a result of their job. TUC research shows that employers who follow the approaches to rehabilitation are far more successful in holding on to employees who are absent for work-related reasons:

- Make rehabilitation a top policy priority.
- Invest in employee health and access to rehabilitation services, such as physiotherapy or counselling.
- Keep in touch with absent employees and offer practical support.
- Be alert to disability issues.
- Do not make health a disciplinary matter.
- Involve line managers in supporting good rehabilitation practice.

It may well help you if your employer responds to your absence at an early stage – six weeks can be too late to begin to offer support. TUC evidence shows that the cost to an employer of providing physiotherapy services is repaid eight times over in the benefits it brings to the employee and the organization by helping the worker get back to work much earlier.

Successful rehabilitation usually involves a **return-to-work agreement** (often in writing) between you and your employer. This may include:

- **lighter or alternative duties**, at least to start with, which have been discussed and agreed with you;
- **a staged return**, eg starting with a few hours or days each week, which you have agreed to;
- **practical details**, eg travel plans, salary level, start and finish times, support from a colleague or mentor at work, training; and
- **a review period**, eg after a few weeks.

The *Employee's Guide* in the *Get Back!* series provides a systematic guide for employees who, due to accident or ill health, are finding it difficult to do their job. It helps you to identify 'difficulties and solutions' with your current job, and offers advice when preparing to meet your employer (see page 229).

Getting advice

When preparing to return to work, you will need advice from your union rep, if available, your GP or other independent adviser. Involving your GP is very important. The GP will have knowledge of specific medical factors that need to be taken into account, which may limit your work activities, for a time or permanently. The GP will also be aware of any other treatment, eg physiotherapy, that may still have to be completed.

GPs may be contacted by an OH doctor or nurse for information on one of their patients who is off sick, or because one of their patients needs specific help with his or her health in relation to work. As we said above, any information provided by a GP is subject to 'patient confidentiality' and to the patient giving prior consent.

Ill health retirement

Ill health retirement may be necessary if you are permanently unable to do the job for which you were employed, and no suitable alternative can be found. Ruth Chambers advises that discussion between the GP and the employer, perhaps through the OH service if there is one, is very important before this decision is taken. The GP will want to take into account the psychological and financial consequences of retirement.

As Chapter 20 shows, permanent disability need not be a bar to future employment. Jobcentres have specially trained disability employment advisers (DEAs), offering advice on:

- alternative work options with an existing employer;
- adaptations of the workstation and workplace, including funding towards costs;
- recommending or providing individuals and firms with special aids and modifications;
- training in any new skills required.

Further information

Chambers, Dr Ruth *et al* (2000) *Occupational Health Matters in General Practice*, Radcliffe Medical Press (see page 248).

Sources

Improving access to occupational health support, *Occupational Health Advisory Committee Report* (2000), available from HSE (tel: 020 7717 6000) or free on HSE's Web site: www.hse.gov.uk/hthdir/noframes/access.htm.

A servant of two masters?, Kloss, Diana (2001) *Occupational Health Review*, **90**, and Phone a friend: employee assistance programmes, *Employee Health Bulletin*, **13**, February 2000, available from tel: 020 8662 2000, price £12 (single copy).

23 Occupational health in the community

Community-based OH projects

In a dozen towns and cities in the UK, including London, Sheffield, Leeds, Manchester, Bradford, Keighley, Liverpool, Rotherham and Lothian, community-based occupational health projects and hazards centres provide specialist advice and support to workers suffering from work-related ill health or injury. The six occupational health projects work in local GP practices. Their free services are an invaluable addition to over-burdened GPs, and include:

- **surgery sessions**: confidential advice sessions in GP surgeries for patients suffering from work-related ill health or injury;
- **advice**: telephone or drop-in advice services;
- **health checks**: including hearing and lung functions tests.

OH advisers will *not* contact your employer without your express consent. You can expect the following support from an OH adviser:

- Confidential advice on the health problems identified.
- An in-depth personal interview about your work, past and present.
- Health checks, if necessary. Some projects provide hearing tests, lung function and breathing tests, checks for peripheral nerve damage and other tests.
- Practical advice on how to deal with your situation. This will cover any new or existing disabilities that affect your work; advice on how to retain your job; benefits and compensation claims; pensions and Statutory Sick Pay entitlement.

- Information on further medical or legal assistance, in consultation with your GP. This may include referrals to other specialist agencies, eg solicitors, physiotherapy, as necessary.
- Advice on contacting your union rep, if there is one at your work, and union legal services.

The projects have built up specialist knowledge about industries and patterns of work-related ill health in their locality. They tailor their services to local needs. A common thread linking occupational health and hazards projects is a desire to 'empower' workers with information, knowledge and confidence about their health condition. As Becky Allen points out in her review of the work of these projects, the goal is 'permanent improvements' in the working conditions of people whom they advise.

The services provided by community-based OH projects are independent from employers and completely confidential. For these reasons, the projects claim that patients prefer their GP-based services to those provided by an employer. Getting independent advice can be crucial to employees when their livelihood depends on their good health.

Support groups

Hazards centres and OH projects also initiate or support local campaigns that focus on the main work-related health issues in the community. These reflect the main industries and occupations in the area. They support a wide range of self-help groups, covering such issues as RSI, homeworking, asbestosis and health for young workers.

Here, we describe some OHPs' services, and contact details.

Where to get help

Sheffield Occupational Health Advisory Service

Established over 20 years ago, Sheffield OH Service was the first of its kind in the UK. Key services include surgery sessions. The service's six advisers offer confidential advice sessions in 24 GP practices across the city. Most of the sessions are in GP surgeries within two areas of the city covered by a health action zone.

The service initially tackled work-related ill health and injuries in Sheffield's traditional manufacturing industries, including such conditions as industrial deafness and vibration white finger, caused by using hand tools. Now, advisers are seeing an increasing number of cases of workplace stress and bullying, from teaching and from 'new economy' sectors such as call centres.

Contact: Sheffield Occupational Health Advisory Service, 55 Queens Street, Sheffield S1 2DX (tel: 0114 275 5760).

Work and health in Sheffield

During a six-month period, the service's advisers interviewed 111 patients referred to them by GPs. Table 23.1 sums up their main occupational health problems, some patients citing more than one health concern. Only a minority of patients had access to a workplace OH service. The project reported 'where they do exist, they are often mistrusted by workers, who regard them as employer-controlled, lacking independence and confidentiality'.

Patients were given advice and information specific to their condition, for example:

- manual handling: advice on employers' duties to eliminate, avoid and reduce hazardous manual handling, and information on good handling techniques;
- chemicals: detailed advice on preventing and coping with dermatitis at work for patients with skin problems;
- VDU users: advice and information on all aspects of the safe use of VDUs, including workstation design, posture and basic ergonomic principles.

Health Works, Newham, East London

Health Works provides occupational health advice sessions to Newham residents. Key services include morning and afternoon surgery sessions in eight GP practices across the borough. Patients are referred from their GP or physiotherapist. Advisers 'signpost' patients to specialist advice, if necessary, eg for counselling,

Table 23.1 Work and health in Sheffield

Work-Related Ill Health Condition	Men	Women
Chemicals and hazardous substances: cleaning agents, hairdressing colours, perms, solvents, detergents, diesel oils, surgical spirit, coolants, acids.	28	23
Noise at work.	24	6
Lifting and manual handling: back problems, strains, sprains. Women mainly from healthcare, catering and shop work. Men from construction, printing, driving, production work.	45	24
Repetitive work and RSI: VDU users, hairdressing, machine operators, shop work, processing.	13	13
Stress Stressors for men – long hours, shiftwork, violence at work, dealing with the public, heavy workload, deadlines, understaffing, poor management. Stressors for women – heavy workload, deadlines, poor management, violence, dealing with the public, long hours.	25	34

(Source: Sheffield Occupational Health Advisory Service)

surgery or to solicitors for employment law advice. Contact is made with employers only with the consent of the patient.

For example, if a patient is suffering from stress, the adviser will talk through the key issues, to help identify the sources of the problem and ways forward. If the stress is work-related, advice may cover ways of raising the issues at work, how to cope with work pressures or, if the patient has resigned, making an Employment Tribunal application.

Specialist projects include homeworking hazards, and work-place safety education for school-leavers.

Contact: Health Works, Alice Billings House, 2–12 West Ham Lane, Stratford, London E15 4SF (tel: 020 8557 6161; e-mail: Healthworks@Newham.gov.uk; Web site: www.healthworks-in-London.org.uk).

Greater Manchester Hazards Centre

The centre provides casework advice, training and support to local workplace and community organizations:

- Telephone and drop-in advice, including advice and representation (eg at Employment Tribunals and for social security appeals), referral to specialist agencies and liaison with local union reps, as appropriate.
- Base for support groups: including Greater Manchester Asbestos Victims Support Group; the Manchester Area RSI Support and Action Group, a self-help group dealing with the 'massive' problem of RSI in the city; and the North West Safety Reps Network, providing support to safety reps in the region.
- Health and safety training and other services, including TUC and trade union courses. Publishes newsletter, *Hazardous Times*, and health and safety factsheets.

Contact: Greater Manchester Hazards Centre, 23 New Mount Street, Manchester M4 4DE (tel: 0161 953 4037; Web site: www.gmhazards.org.uk).

Leeds Occupational Health Project

Key services include:

- Surgery sessions. Patients waiting to see the GP or practice nurse are offered a health interview about their work, past and present. Occupational history and exposure to workplace hazards then become part of the patient's medical record held by the GP. Workplace health advice is given to around 80 per cent of patients seen. Advisers provide further support, or referral to a specialist in the field. Advisers also offer to help patients to liaise with their union rep, as appropriate.
- Telephone and drop-in advice: at the centre in central Leeds.
- Health surveillance: hearing tests and lung function tests.
- Public resource centre: providing advice and information on all aspects of work-related health.

GP satisfaction

Since the Leeds OH Project began offering GP surgery sessions in 1993, it has seen more than 10,000 patients in 36 different practices across the city. The project's annual report for 2000 shows that work-related stress is now the biggest single health issue presented by patients. The main reasons include increasing workloads and a greater willingness on the part of men to admit to stress, and see their GP about it.

Case study

A 57-year-old man, suffering both dizziness and light-headedness, was referred by his GP. The patient was employed as a painter and shotblaster. It was suspected that his condition was a result of exposure to solvents in paint. The OH adviser carried out lung function tests, and concluded that the breathing apparatus provided by the employer was unsuitable for paint spraying. The patient was provided with written advice specifying that the employer had a duty to provide equipment that met the requisite health and safety standards for carrying out this type of work.

Contact: Leeds Occupational Health Project, 88 North Street, Leeds LS2 7PN (tel: 0113 294 8222; Web site: www.leedsohas.org.uk).

London Hazards Centre

The centre provides advice, information, training and other forms of practical support to Londoners concerned to protect their health and safety at work:

- Casework. The telephone advice line (or drop-in) for casework includes first-stage legal and medical advice, and assistance in contacting other specialist agencies, your GP or a trade union.
- Advice and information to workplace and community groups.
- Factsheets: from asbestos to violence.
- Training. One-day courses include workplace safety management, risk assessment, VDU hazards, tackling stress at work.
- *Daily Hazard*: quarterly newsletter, on subscription.

Contact: London Hazards Centre, Hampstead Town Hall Centre, 213 Haverstock Hill, London NW3 4QP (tel: 020 7794 5999; Web site: www.lhc.org.uk).

Funding community-based occupational health

Funding for community-based occupational health projects is insecure, and based on short-term grants from the Lottery, health action zones, the local council, NHS trusts and Government regeneration programmes. The Occupational Health Advisory Committee recommended in 2000 expanding the number of projects, and providing 'secure funding for existing projects to allow evaluation'. It said that heavily-burdened GP practices, even where there are doctors or nurses with occupational health experience, 'are unlikely to remain continuously focused on the work-related health needs of their patients'. To date, no national funding is available, despite the proven effectiveness of community-based OH services, although the HSE has launched a two-year study of their contribution to improving occupational health.

Further information

Allen, Becky (2001) OHPs come of age, *Occupational Health Review*, **90**, available from tel: 020 8662 2000, price £12 (single copy).

For a full list of local hazards campaigns and occupational health projects, contact Hazards Campaign (tel: 0161 953 4037; Web site: www.hazards.org).

Further contacts:

- Bradford Workers Health Advice Team, WHAT, Unison Building, 2nd Floor, Auburn House, Upper Piccadilly, Bradford BD1 3NU (tel: 01274 393949).
- Keighley Work Safe Project, 136 Malsis Road, Keighley BD21 1RF (tel: 01535 691264; Web site: www.worksafe.org.uk).
- Liverpool Occupational Health Advisory Service, Melbourne Buildings, 21 North John Street, Liverpool LS2 5QU (tel: 0151 236 6608; e-mail: liverpool-ohp@fsmail.net).
- Lothian Trade Union and Community Resource Centre, and Lothian RSI Support Group, 26/28 Albany Street, Edinburgh EH1 3QH (tel: 0131 556 7318; e-mail: ltucrc@aol.com).
- Rotherham Occupational Health Advisory Service, Room 9, Imperial Buildings, Corporation Street, Rotherham S60 1PA (tel: 01709 820472; e-mail: ohasrpct@tiscali.co.uk).

24 Making a personal injury claim

In the nightmare scenario that every bank and post office employee dreads, the branch of the high street bank where Susan Macey worked was raided. Susan, who worked at the front of the bank with no protection, was held with a gun to her head.

After the incident, instead of offering her support, the bank demanded that Susan returned to her job the following day. Her requests to be moved to a back-office job were ignored. Her employer failed to deliver the counselling they offered her.

Susan began to suffer panic attacks, and felt under extreme stress at work. Although she was eventually offered a transfer to a back-office job, she decided to take medical retirement. Her union, Unifi, took up her case and covered the costs of a claim to the Criminal Injuries Compensation Authority, where she won compensation of £20,000.

In the 10 years to 2001, unions won more than £3 billion in compensation payments for their members who were victims of work-related injury or ill health. TUC figures show that unions take on more than 50,000 cases each year, with an average settlement exceeding £6,000 per employee. The six-figure settlements that hit the headlines are uncommon. They usually result from an employee facing many years of lost earnings in a well-paid job, following a calamitous or life-threatening injury or illness.

Yet, the TUC points out, 'Unions don't want as many cases as they are getting. These cases are a sign that too often there are inadequate health and safety checks. And, in too many cases, victims are offered no rehabilitation to get them back to fitness and back to work'.

Unions report an increase in cases involving stress, RSI, asbestos and vibration white finger.

Claiming for personal injury

A work-related 'personal injury' includes any 'disease, impairment of physical and mental condition, and death' caused, or made worse by, your employment.

If you are injured at work or in connection with your work, or suffer ill health as a result of the job you do, you may be entitled to recover from your employer compensation for the injury itself ('pain and suffering') and for financial losses and expenses caused by the injury. Here, we explain what is involved in making a claim, whether or not you are a union member. Claiming is considerably easier if you are in a union. But, in any event, you will need special legal advice. The TUC's guide, *Your Rights at Work,* provides a useful introduction to the basis of law at work.

Your case

You must be able to prove it is more likely than not that:

- you have been **injured** (see the definition above);
- your employers **failed in their legal duty of care** towards you;
- your injury was **caused** by that failure, not by some other factor, eg an underlying medical condition;
- it was **foreseeable**, on the information and knowledge available at the time of the employer's failure, that the failure would lead to your injury.

Time limit

You must bring your claim by starting legal proceedings within three years of the date of the accident or (in disease cases) of the date when you first realized that you had suffered some kind of injury caused by your employer. That 'date of knowledge' will often be long before a doctor gives you a definite diagnosis, so don't delay in starting a claim. You should seek legal advice as soon as possible so that evidence can be preserved, witnesses contacted, photographs taken, etc.

What are the employer's obligations?

Your employer, its subcontractors, its main contractor and the providers of work equipment owe *legal obligations* to you. Some of these obligations have been created by judges, and some have been made by Acts of Parliament or Regulations. Some of them are (or should be) enforced by the state through the inspectors of the HSE or the environmental health officers of local authorities. Some of them can be asserted by you as the basis for a compensation claim.

The main obligations of the employer are to provide:

- safe premises;
- suitable, properly maintained equipment;
- a safe system of work, and 'safe' people to work with;
- adequate training, information, instruction and supervision;
- suitable and sufficient *assessments of risks*, and action to *eliminate* or *minimize* risks that are identified, including further risk assessment if the circumstances change;
- suitable personal protective equipment.

The enforcing authorities can issue improvement or prohibition notices and can prosecute. The Human Rights Act 2000 offers potential new ways for persuading the authorities to take action when they seem reluctant to do so. However, their involvement is not essential for you to make a compensation claim.

Which courts?

If you have been injured, you can bring a private claim for compensation against your employer. The state does not help you to do this; you have to seek your remedy as a private individual. See below for advice on finding a legal representative to help you do this. Your claim will usually be decided in the County Court if it is worth about or below £50,000, or in the High Court if it is substantially above £50,000. However, Employment Tribunals may also have jurisdiction to give compensation for personal injury in some cases, eg under the Race Relations Act where psychiatric injury such as depression has been caused by racial harassment.

Specialist legal advice

You will need *specialist legal advice* to make your claim, whether or not it is settled out of court, or goes to court.

The following points apply if you are a union member:

- Unions provide free legal advice and representation to their members in personal injury cases. Each union follows its own procedure, so you will need to enquire. Usually, individual members access their legal services through their workplace union branch, their nearest district office (details in the membership card) or head office legal service. On page 271, we list union telephone helplines providing advice on personal injury and other issues.
- Most unions use a questionnaire or application form, asking for basic information on:
  - the date, time and place of the injury or illness;
  - a brief description of the injury;
  - whether the injury or illness was recorded in an accident book, or reported to management;
  - details of any witnesses, etc;
  - other medical information via your GP;
  - absences from work; and
  - a first estimate of your financial losses.
- The questionnaire may be returned direct to the union's legal department or their solicitors, to a local branch or to a full-time officer.
- Solicitors retained by unions will operate a 'merits' test on the cases they handle. This varies between unions, with some prepared to take a greater risk in 'high-profile' cases that appear to break new ground, eg stress cases.
- Your union solicitor should keep you informed at all stages in the progress of your case, including arranging for expert opinions (eg medical opinions). At some stage, they may advise you that your employer is willing to settle out of court. Here, you will have to take heed of this advice, and make your own assessment of the risk of proceeding to court or accepting

an out-of-court award. Remember, the average wait for a case to be settled is three years.

These points apply if you are not a union member:

- **Check your insurance policies**. Your home contents insurance, motor insurance or credit card protection may include legal expenses insurance. This may give you free legal advice and representation, subject to the merits of your claim and the insurer's indemnity limit under the policy. You are entitled to assert your own choice of lawyer, although your insurers may want you to use one of their panel solicitors.
- **Contact your nearest CAB or law centre**. Details are in the phone book. Your CAB or law centre will help you to identify the nature and likely cause of the injury, and the possible remedies open to you. If you are a union member, they will encourage you to report the injury, if you have not already done so, and to contact your union representative. They will refer you to a specialist personal injury lawyer with the right 'quality marks'.
- **Specialist personal injury lawyers**. These include members of the Association of Personal Injury Lawyers (tel: 0115 958 0585), members of the College of Personal Injury Law, those accredited by the Law Society's Personal Injury Panel, and a legal aid franchise (Law Society, tel: 020 7242 1222).
- **Beware of TV and other adverts**. 'Claims farmers', claims assessors and the like may seem to offer a convenient, friendly and trouble-free way to start your claim. Some can offer this service, but think of them as door-to-door sales reps and 'middlemen'. And watch out for the small print of any agreement they may want you to sign. Ask them what you stand to gain out of the agreement, and what they will gain. Ask if the person you are talking to is a qualified lawyer. Then check what they are offering you, by telephoning a firm of solicitors who have the 'quality marks' mentioned above, before you sign an agreement.

How does the legal system work?

Litigation is expensive, financially and emotionally. For union members, the union bears the full cost of the case. Meanwhile, union member or not, you may have to prove every issue in your claim, from the way in which your accident happened to the full effects of your injuries. The whole of your past medical and psychiatric history is likely to be picked over and commented on. You may need the evidence of experts to help you prove your claim, and in a case of serious injuries these may include engineers, medical experts, occupational therapists, employment consultants and barristers. This work has to be financed, and few people could afford to fund it from their own resources.

Legal aid is not available for the great majority of personal injury claims from April 2000. The Government has decided that it is no longer required for most compensation and other money claims, and hopes that law firms and insurance companies will be willing to absorb the costs and financial risks of personal injury litigation.

The alternatives to legal aid are:

- lawyers retained by your union, or other trade or professional body of which you are a member;
- lawyers paid by 'before the event' legal expenses insurance;
- lawyers who offer you a 'no win, no fee' agreement, usually backed up with:
  - an insurance policy to protect you against liability for your opponent's legal costs (and sometimes your own legal expenses) if you lose your claim;
  - a loan agreement to pay for legal out-of-pocket expenses, such as experts' fees.

What your solicitor will decide with you

Most solicitors nowadays offer an initial meeting or telephone consultation without obligation, at the end of which they should send you a detailed letter stating:

- what they think are your prospects of success;
- what the likely outcome in compensation will be;

- what the total legal costs are likely to be;
- on what terms they propose to fund the case, including:
  - who will have responsibility for the conduct of the case (and their level of qualification);
  - what you will be charged for their work; and
  - if a 'no win, no fee' agreement is offered, what the success fee will be, who will pay it, what arrangements will be made for a loan for expenses and which insurance company will be invited to offer you an indemnity for legal costs.

How to deal with offers of settlement

At strategic points in the preparation of your case, your solicitor may advise you that an offer of settlement should be made to your opponent. This may be a 'global figure' for your whole claim, or it may be an offer on just one or more of the arguments in the case, for example on the question of how the responsibility for the accident is to be apportioned, or how much your compensation for the injury itself should be.

The purpose of making the offer is to speed up the claim and reduce your legal costs. The jargon for this procedure is 'making a Part 36 offer', and your solicitor should ask for your confirmation that you agree to it being made. The offer remains open for 21 days from the date when your opponent receives it. If your opponent refuses the offer, you press on to trial and the court's decision is better than your offer, your opponent will have to pay a high rate of interest on your damages and extra legal costs. Of course your opponent may also make a Part 36 offer, backed up with a payment into court, which puts you in the position of deciding whether to take a gamble on beating their offer (this is discussed elsewhere in this chapter).

Remember that the compensation is for you, and that you are entitled to know, and need to know, what deductions if any will be made from your compensation.

What happens if I win?

You may win by agreement (ie an out-of-court settlement), or as a result of a court hearing. The great majority of personal injury claims are settled before they go to a full court hearing. But you are unlikely to get the best outcome unless you are prepared to go to court if necessary, and your opponent knows that.

If you win, your opponent pays:

- **damages**, or compensation, usually assessed under two main categories:
  - **'general damages'**, which is a figure worked out from a little paperback book for judges, backed up by case reports on similar cases, for your pain, suffering and loss of quality of life (short-term or long-term);
  - **'special damages'**, which is the total of the financial losses and expenses (including those that you are likely to incur in the future) resulting from your injuries (the list may be long, and it may include not only obvious items like loss of earnings but also, in cases of catastrophic injuries, the cost of adapting your home to cope with your disability, or your care needs for the rest of your life);
- **interest on damages**;
- **your legal costs** (subject to an assessment of the amount, by the court, if not agreed).

What happens if I lose?

'Losing' can mean not only failing to prove one of the four elements of a personal injury claim listed above, eg that your opponent failed in their duty of care towards you. It can also mean failing to beat a payment made into court by your employer, or the offer of settlement made by your opponent, before the hearing. The effects are: **if you fail to prove your case**, you get no damages or compensation and you are liable to pay your opponent's legal costs, which is why you may need legal expenses insurance; or, **if you fail to beat a payment into court**, you pay all your costs *and* all your opponent's costs from the date of their payment in until the end of the case: when you consider that the costs of a trial are probably 25 to 30 per cent of the total costs of the case, you will see that failure to beat a payment into court made strategically close to the trial can wipe out the amount of the compensation even in a big-value case.

How long before I receive compensation?

The legal rules on personal injury claims were changed in 1999 to try to speed up the progress of claims. A set of pre-trial rules, known as a 'protocol', was established, which requires your legal representative to put sufficient detail in a letter of claim so as to enable the defendant (your employer) to investigate your claim. The defendant has to acknowledge the letter and then has three months to decide whether or not to admit responsibility for the injury. During that time your representative will obtain your medical records and a medical report.

If you have to start a court case, the court will hold a case management conference and set a timetable for the remaining stages of preparation of the case and a date for the trial. The time taken for this work varies for each individual case, and depends on the extent to which the parties can agree the main issues and the evidence. Some cases may be settled in a matter of months. Others may take two to three years to come to conclude, particularly where the claimant's medical condition has not stabilized. However, in many cases the claimant will be entitled to request one or more interim payments of compensation.

Evidence needed

To strengthen your case, you should do the following:

- Start gathering evidence straight away and keep a dossier of:
  - your notes of how your accident/exposure to hazardous substances etc happened;
  - sketch plans, photos, details of the make and manufacturer of relevant equipment;
  - names, addresses, phone numbers of witnesses;
  - a diary of your symptoms, and medical treatment;
  - receipts for all expenses (even tickets for travelling to and from hospital);
  - payslips before and after the accident.

- Make sure the accident is recorded at work and, if relevant, that you have informed the Jobcentre.
- Seek medical treatment promptly and always attend medical appointments (not attending may be interpreted as meaning that your condition was not very serious) or rearrange them well in advance, but don't discuss your compensation claim with your doctor because all your medical records will be disclosable to your opponent.
- Keep your legal representative informed of developments, including any disciplinary procedures brought against you, and any changes in your pay or conditions following a return to work.
- Get advice on benefits.

Further information

For details of your nearest law centre, you can ring the Law Centres Federation tel: 020 7387 8570; or go to www.lawcentres.org.

25 Union advice and support

Safety studies reveal that serious accident and ill health levels are lower in unionized workplaces. From their earliest days, trade unions have treated health and safety at work as a core concern. Today, they bring a wealth of experience and knowledge to workplace safety issues. They offer expertise and legal support to those facing work-related ill health or injury.

- **Union dues**. It costs less than you may think. Weekly union dues cost under £1 for part-time workers and around £2 if you work full time. Joining a union is confidential – contact is not made with your employer.
- **Joining a union**. To find out more about how to join a union, and which one is the most appropriate for you, you can use the TUC's online union finder at www.worksmart.org.uk/unionfinder. This includes a directory of every union affiliated to the TUC. Alternatively you can phone the TUC's Know Your Rights line on 0870 600 4882 (calls charged at national rate).
- **Oneline help and advice**. The TUC's world of work Web site workSMART (at www.worksmart.org.uk) has hundreds of questions and answers on topical work issues. workSMART includes a health section that covers many of the subjects covered in this book and a more general guide to your rights at work. It includes links through to other useful and relevant Web resources.
- **TUC's Know Your Rights Line**. Whether you are in a union or not, the TUC's helpline on 0870 600 4882 (calls charged at national rate) can send you a range of leaflets covering a wide range of work related issues. These include a wide range of rights issues, and leaflets relevant to the content of this book on bullying, violence at work and safety for younger workers.

What to expect from a safety rep

The Safety Representatives and Safety Committees Regulations 1977 give recognized trade unions the right to appoint workplace safety reps. The Regulations require employers to set up a safety committee. Safety reps have the right to play an active part in workplace risk assessments, investigate hazards and 'dangerous occurrences', and investigate members' complaints (see page 33). Many safety reps have been trained in health and safety at work.

Unions advise their safety reps to:

- Treat any cases seriously, and be supportive. Ongoing support may be essential if bullying or harassment is involved. In many such cases it will be the word of the bully or harasser, who is more often in a more influential position, against the word of the target.
- Listen carefully to what the employee says. Show that you are prepared to help.
- Encourage the worker to write down details of each incident, including what was said and done, and the date and time.
- Advise the employee to report the incident in the accident book.
- Find out whether other workers have experienced similar problems, and if so ask them for details.
- Discuss how the employee wants the case to be pursued. Describe the stages in any procedure that you may follow.
- Offer to represent and/or support the employee at any meetings with management.
- Take any documentation needed with you. Discuss the case beforehand, and keep the employee informed at all stages. Make sure that any possible solutions are discussed and agreed with the employee.
- If necessary, seek help from the union's full-time officer, legal service or health and safety specialists.
- If the worker agrees, seek the help of other employees/union workers.
- Ensure that the case is dealt with as quickly as possible.
- Ensure that any agreement reached is effective.

(Adapted: Amicus)

Union helplines

Some unions now operate helplines, listed in Table 25.1. Alternatively, contact the unions direct, through their national or local offices (see Useful addresses).

The right to be accompanied at work

If you are tackling a *health and safety issue* at work, you now have the right to be accompanied at meetings with management. The Employment Relations Act 1999 introduced a new right for workers to be accompanied at a disciplinary or grievance hearing by a trade union representative or a fellow worker. A worker who is 'required or invited' by his or her employer to attend a disciplinary or grievance hearing may 'reasonably request' to be accompanied by a 'single companion' (Employment Relations Act 1999, s 10).

Your companion can be either a paid trade union officer, a lay trade union official or a fellow worker:

Table 25.1 Examples of union helplines

Union	Helpline Number
CWU	020 8971 7380/7593
Fire Brigades Union	0800 783 4778
GMB health and safety unit	020 8947 3131
NASUWT teachers' 24-hour stress counselling helpline	0800 056 2561
NUMAST Care	01455 890055
PCS accident claims line	020 7801 2651
PCS domestic violence helpline	0345 023 468
PCS health and stress helpline	0870 523 4998
PCS racial incident helpline	020 7801 2678
Prospect safety line	020 7902 6635
TGWU CareXpress legal helpline (or contact nearest local office)	0800 709007
TSSA members' helpline	0800 328 2673
TUC Know Your Rights line	0870 600 4882
Unifi legal helpline	0870 755 2233
Unison Direct	0800 597 9750

- All 'workers' are covered by the new rights. 'Worker' includes homeworkers, agency workers, part-time, temporary and casual workers, those on short-term contracts and those who work overseas.
- The statutory right to be accompanied at meetings with your employer applies to any hearing connected with 'the performance of a duty by the employer in relation to a worker'. Because *health and safety* involves your employer's *legal and contractual* obligations, you have the right to be accompanied at meetings or hearings dealing with health and safety issues.

The choice of companion lies entirely with the worker and includes a paid official of a trade union (regardless of whether the union is recognized or has other members in the workplace); a shop steward who has experience of, or training in, acting as a worker's companion at disciplinary or grievance hearings; and a fellow worker.

These rights are set out in a code of practice issued by The Government's Advisory, Conciliation and Arbitration Service (ACAS). Its Code of Practice on Disciplinary and Grievance Procedures outlines best practice and provides a model approach on the new procedures.

The revised Code of Practice is available on the ACAS Web site at: www.acas.org.uk, or tel: 08457 474747.

Further information

The Employment Relations Act 1999: A guide for trade unionists, November 1999, and *Disciplinary and Grievance Procedures: A guide to the new law*, September 2000, Labour Research Department, 78 Blackfriars Road, London SE1 8HF (tel: 020 7928 3649; Web site: http://www.lrd.org.uk/).

Statutory Recognition: A guide, UNISON, June 2000. Available by calling UnisonDirect (tel: 0800 597 9750, opening hours – weekdays, 6 am to midnight, and Saturday, 9 am to 4 pm; e-mail: direct@unison.co.uk).

Taking Your Time, UNISON guide to time off and trade union facilities (stock number 1608), UNISON (Web site: http://www.unison.org.uk/).

Index

Index of advertisers